BrightRED Study Guide

Curriculum for Excellence

N5

CHEMISTRY

Shona Scheuerl, Robert West and Shona Wallace

First published in 2013 by:
Bright Red Publishing Ltd
1 Torphichen Street
Edinburgh
EH3 8HX

Reprinted with corrections 2013, 2015 and 2016

New edition published 2018. Reprinted with corrections 2018

A CIP record for this book is available from the British Library

ISBN 978-1-906736-95-8

With thanks to:
PDQ Digital Media Solutions Ltd, Bungay (layout) and Patrick Fox (copy-edit)
Cover design by Caleb Rutherford – eidetic

Acknowledgements
Every effort has been made to seek all copyright-holders. If any have been overlooked, then Bright Red Publishing will be delighted to make the necessary arrangements.

Permission has been sought from all relevant copyright holders and Bright Red Publishing are grateful for the use of the following:
Les Chatfield/Creative Commons (CC BY 2.0)[1] (p 6); Michael Welsing/Creative Commons (CC BY-ND 2.0)[2] (p 6); Alchemist-hp/Creative Commons (CC BY 3.0)[3] (p 13); H. Michael Miley/Creative Commons (CC BY-SA 2.0)[4] (p 13); Steve Jurvetson/Creative Commons (CC BY 2.0)[1] (pp 14 and 26); Hey Paul/Creative Commons (CC BY 2.0)[1] (p 15); Mike Baird/Creative Commons (CC BY 2.0)[1] (p 26); GorillaGolfBlog/Creative Commons (CC BY-SA 2.0)[4] (p 27); Eugen Lehle/Creative Commons (CC BY-SA 3.0)[5] (p 36); Lee Pullen (p 38); NASA/George Shelton (p 40); Two photos by Robert West (p 50); Denis Tabler/Shutterstock.com (p 62); Africa Studio/Shutterstock.com (p 62); Andrey Starostin/Shutterstock.com (p 65); Nomad_Soul/Shutterstock.com (p 65); stormdance/StockXchng (p 66); tristan tan/Shutterstock.com (p 66); farbled/Shutterstock.com (p 74); Dino Osmic/Shutterstock.com (p 76); Ian Barbour/Creative Commons (CC BY-SA 2.0)[4] (p 83); Ilya Rabkin/Shutterstock.com (p 91); Bork/Shutterstock.com (p 94); wenht/istockphoto (p 94). Bright Red Publishing would also like to thank the Scottish Qualifications Authority for use of Past Exam Questions: SQA Standard Grade Credit Chemistry 2001 Paper (p 107) and SQA Intermediate 2 Chemistry 2012 Paper (p 109). Answers do not emanate from the SQA.

[1] (CC BY 2.0) http://creativecommons.org/licenses/by/2.0/
[2] (CC BY-ND 2.0) http://creativecommons.org/licenses/by-nd/2.0/
[3] (CC BY 3.0) http://creativecommons.org/licenses/by/3.0/
[4] (CC BY-SA 2.0) http://creativecommons.org/licenses/by-sa/2.0/
[5] (CC BY-SA 3.0) http://creativecommons.org/licenses/by-sa/3.0/

Bright red Publishing would also like to thank the Scottish Qualifications Authority for the use of Past Exam Questions. Answers do not emanate from SQA.

Cover image © Caleb Rutherford

Printed and bound in the UK.

CONTENTS

BRIGHTRED STUDY GUIDE: NATIONAL 5 CHEMISTRY

Introducing National 5 Chemistry 4

1 CHEMICAL CHANGES AND STRUCTURE

Reaction rates: Measuring and calculating rates 6

Reaction rates: Graphing the rate of reaction 1 8

Reaction rates: Graphing the rate of reaction 2 10

Atomic structure: The Periodic Table 12

Atomic structure: The structure of the atom 14

Atomic structure: Nuclide notation of
atoms and ions 16

Atomic structure: Isotopes and relative atomic mass 18

Chemical bonding: Ionic bonding 1 20

Chemical bonding: Ionic bonding 2 22

Chemical bonding: Covalent bonding 1 24

Chemical bonding: Covalent bonding 2 26

Formulae and reaction quantities:
Chemical formulae 28

Formulae and reaction quantities:
Writing formulae of compounds 1 30

Formulae and reaction quantities:
Writing formulae of compounds 2 32

Formulae and reaction quantities:
The mole 1 – gram formula mass 34

Formulae and reaction quantities:
The mole 2 – concentration 36

Formulae and reaction quantities:
The mole 3 – equations 38

Formulae and reaction quantities:
The mole 4 – equations and calculations 40

Formulae and reaction quantities:
The mole 5 – percentage composition 42

Acids and alkalis 44

Acids and bases: Making acids and alkalis 46

Acids and bases: Neutralisation reactions 48

Acids and bases: Titrations 50

2 NATURE'S CHEMISTRY

Homologous series: Alkanes 52

Homologous series: Alkenes 54

Homologous series: More hydrocarbons 56

Homologous series: Isomers 58

Homologous series: Hydrocarbon reactions 60

Everyday consumer products 62

Everyday consumer products: Carboxylic acids 64

Energy from fuels 1 66

Energy from fuels 2 68

3 CHEMISTRY IN SOCIETY

Metals: Bonding and properties 70

Metals: Reactions of metals 72

Metals: Extraction of metals 74

Metals: Electrochemistry 1 76

Metals: Electrochemistry 2 78

Plastics, polymers and polymerisation 80

Fertilisers: Commercial production 82

Fertilisers: Making fertilisers 1 84

Fertilisers: Making fertilisers 2 86

Nuclear chemistry: Radioactivity and properties . . . 88

Nuclear chemistry: Radioactive decay and half-life 90

Nuclear chemistry: Half-life and carbon dating 92

Nuclear chemistry: Radioisotopes and uses 94

4 SKILLS

Everyday laboratory skills 96

Standard solutions 98

Making a salt 100

Tables, bar charts and graphs 1 102

Tables, bar charts and graphs 2 104

Analysing data from tables, bar charts and
graphs 1 106

Analysing data from tables, bar charts and
graphs 2 108

Experimental design 110

GLOSSARY

GLOSSARY 112

INTRODUCING NATIONAL 5 CHEMISTRY

This course allows you to develop a wide range of life and scientific skills that will equip you for a future of changing challenges. Its structure provides you with opportunities to develop and extend a wide range of chemistry-focused skills, while helping you to develop an understanding of chemistry's role in the scientific issues that affect society.

THE BENEFITS OF NATIONAL 5 CHEMISTRY

The course is split into three units, each with a real-life theme. The course content will help you to develop knowledge and understanding of the chemistry around you. There will be plenty of opportunity to develop skills of scientific inquiry, as well as investigative and analytical thinking skills, within a chemistry context. Experimental work will allow development of planning and practical skills as well as building an awareness of safety considerations.

The National 5 course is a way to greatly enhance your understanding of the chemistry affecting your everyday life, while developing skills that will help you unravel the scientific issues affecting society and support you in many aspects of your life.

There is quite a lot to come to terms with in this course but, broken down as it is here, it is fairly straightforward. So, what is the structure?

THE EXTERNAL ASSESSMENT

At the end of the course you will be assessed externally by two components:

Component 1 – Question Paper (80% of total mark)

This involves a two-hour question paper in which:

- 25 marks are allocated to an objective test
- 75 marks are allocated to the written paper, which will include questions requiring a mixture of short (restricted) and extended answers.

The majority of marks are given for demonstrating and applying knowledge and understanding. The other marks will be given for applying scientific inquiry, analytical thinking skills.

The question paper will sample skills, knowledge and understanding of the key areas listed below.

The key areas are:

- rates of reaction
- atomic structure and bonding related to properties of materials
- formulae and reacting quantities
- acids and bases
- systematic carbon chemistry
- everyday consumer products
- energy from fuels
- metals
- plastics
- fertilisers
- nuclear chemistry
- chemical analysis.

These key areas may be grouped together into three units: Chemical Changes and Structure, Nature's Chemistry and Chemistry in Society.

In addition, there will be two open-ended questions in the paper. Each question will be awarded 3 marks and can be recognised by the phrase 'using your knowledge of chemistry'. The question will not directly assess knowledge taught during the course. Instead you are to

cont

use the knowledge you do have to suggest possible answers. There is no correct answer and marks will be awarded according to whether you have shown that you have a 'good' (3 marks), 'reasonable' (2 marks) or 'limited' (1 mark) understanding of the chemistry in the question.

A data booklet containing relevant data and formulae will be provided.

The question paper will be written and marked by the Scottish Qualifications Authority (SQA).

Component 2 – Assignment (20% of total mark)

The assignment will be an in-depth study of a chemistry topic chosen by you. You will investigate and research the underlying chemistry of the topic as well as the impact the topic has on society or the environment.

There will be 20 marks awarded for the assignment (and this will be scaled to 25) and the majority of these will be awarded for applying scientific inquiry and analytical thinking skills. The other marks will be awarded for applying knowledge and understanding relating to the topic.

The assignment will assess skills such as handling and processing of data gathered from an experiment and research data that cannot be assessed through the question paper.

There are two stages to the assignment.

First stage – Research

This will be carried out under some supervision and control and during this stage you will:

- agree a topic with your teacher
- agree an aim – you will be given advice to help you choose a suitable aim
- carry out an experiment which allows measurements to be made
- undertake research using websites, journals and/or books to provide data and information that can be used to compare with your experimental results.

Second stage – Report

This will be carried out under a high degree of supervision and control and a maximum of 1 hour and 30 minutes is allowed. It is marked externally by SQA. Your report should include:

- an aim
- chemistry knowledge and understanding relating to the topic
- a description of the experiment
- data from your experiment – both raw and processed, including graphical representation
- data from an internet or literature source along with a reference
- analysis of your results
- conclusion
- evaluation.

During the report stage you will only be allowed to take in:

- the Instructions for Candidates
- your raw experimental data
- the internet or literature data
- information on the underlying chemistry
- the experimental methods.

HOW WILL THIS GUIDE HELP YOU MEET THE CHALLENGES?

The aim of this book is to help you achieve success in the final exam by providing you with a concise coverage of the key areas of the course. Helpful hints are provided throughout the book in the 'Don't forget' features, while there are plenty of opportunities to practise applying your knowledge through 'Things to do and think about' and the online tests. Some of the skills you will be expected to demonstrate are also covered in the book. These may be in with the relevant key areas or covered in separate sections.

REACTION RATES: MEASURING AND CALCULATING RATES

Iron nails rusting – a slow reaction

Chemical reactions can take place at different speeds or rates. Some reactions, like methane burning, are very fast. Other reactions, like iron rusting, are slow.

The rate of a chemical reaction can be altered by changing the reaction conditions. In general, the rate of a reaction can be increased by a rise in temperature, an increase in reactant concentration or an increase in the surface area of a solid reactant.

Catalysts, which will be discussed in the Chemistry in society chapter, also increase the rate of a reaction.

An explosion – a very fast reaction

HOW TO MEASURE REACTION RATE

During a chemical reaction the reactants change into products. This means that as the reaction proceeds the reactants are being used up and the products are being formed.

There are several methods of monitoring the rate of a reaction. They all rely on measuring how much of a reactant is used up or how much of a product is formed in a given period of time.

The reaction between marble chips (a form of calcium carbonate, $CaCO_3$) and dilute hydrochloric acid, HCl, is useful to show how the rate of a reaction can be measured when a reaction produces a gas.

The **balanced formula equation** for the reaction is

$$CaCO_3(s) + 2HCl(aq) \rightarrow CaCl_2(aq) + CO_2(g) + H_2O(l)$$

In this reaction, the gas carbon dioxide, CO_2, is produced. If appropriate apparatus is used, the mass or volume of the gas can be measured as the reaction proceeds.

Measuring the volume of gas produced

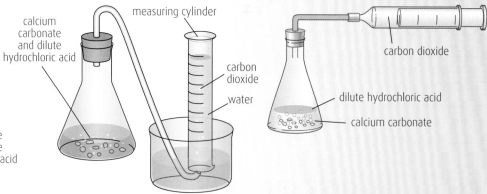

Two methods of measuring the volume of gas produced in this reaction

Using this apparatus the volume of carbon dioxide produced is measured with the syringe or the measuring cylinder at suitable time intervals.

Measuring the mass of gas produced

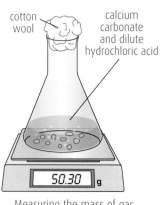

Measuring the mass of gas produced in this reaction

Using this apparatus the carbon dioxide gas produced will escape from the flask and the mass of the flask and its contents will decrease. The change in mass, which is equivalent to the mass of carbon dioxide produced, can be determined at suitable time intervals. The loose plug of cotton wool will not prevent the gas from escaping but it will stop any acid spray escaping as the mixture fizzes.

A FORMULA FOR REACTION RATE

The results from experiments similar to those outlined on page 6 can be used to calculate the reaction rate.

The rate of a reaction can be determined using the relationship

reaction rate = $\frac{\text{change in quantity of reactant or product}}{\text{change in time}}$

EXAMPLE

During a reaction, 25 cm³ of gas were produced in the first 50 seconds. Calculate the average rate of reaction for the first 50 seconds.

reaction rate = $\frac{25}{50}$ = 0·50 cubic centimetres per second (cm³ s⁻¹)

EXAMPLE

Aiden monitored the rate of the reaction between calcium carbonate and dilute hydrochloric acid. He measured the volume of gas produced as the reaction progressed. Aiden's results are shown in the table.

Time(s)	Volume of carbon dioxide (cm³)
0	0
5	30
10	60
15	83
20	100
25	100

Calculate the average rate, in cm³ s⁻¹, at which carbon dioxide is produced between 5 seconds and 10 seconds.

rate = $\frac{60 - 30}{10 - 5}$ = $\frac{30}{5}$ = 6·0 cm³ s⁻¹

 THINGS TO DO AND THINK ABOUT

1   What is meant by **reaction rate**?

2   The apparatus shown was used to monitor the rate of the reaction between magnesium and dilute sulfuric acid.
The volume of hydrogen gas produced was measured for 5 minutes.

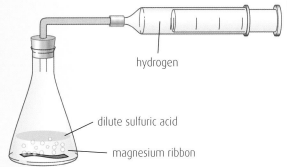

hydrogen

dilute sulfuric acid

magnesium ribbon

The results are shown in the following table:

Time (min)	Volume of hydrogen (cm³)
0	0
1	40·0
2	80·0
3	110·0
4	X
5	132·5

(a) Calculate the average rate of the reaction, in cm³ min⁻¹, between 1 and 3 minutes.

(b) Deduce the value of volume X.

REACTION RATES: GRAPHING THE RATE OF REACTION 1

When an investigation into reaction rates is carried out, the experimental results can be displayed on a graph. The graph helps to visualise what happened over the course of the reaction.

THE IMPORTANCE OF UNITS

The concept of units is crucial when measuring any given quantity in chemistry. A numerical answer without a unit tells us nothing about what has actually been measured. Consider the phrase 'I measured out 20 of water'. 20 what? It is much clearer to report 'I measured out 20 g of water', as this correctly identifies that it was the mass of water being measured.

Changes in mass

If a gas is produced in a reaction and the changes in mass are recorded at regular intervals then

$$\text{reaction rate} = \frac{\text{change in mass}}{\text{change in time}}$$

If the time is measured in seconds, the relationship for units will be

$$\text{reaction rate} = \frac{g}{s}$$

The unit for reaction rate is grams per second and this is written as $g\,s^{-1}$.

Changes in volume

If the volume of gas produced is measured at regular intervals, then

$$\text{reaction rate} = \frac{\text{change in volume}}{\text{change in time}}$$

If the time is measured in minutes, the relationship for units will be

$$\text{reaction rate} = \frac{cm^3}{min}$$

The unit for reaction rate is cubic centimetres per minute and this is written as $cm^3\,min^{-1}$.

ONLINE

Click the 'How to draw graphs' link at www.brightredbooks.net/N5Chemistry.

WHAT DO REACTION RATE GRAPHS TELL US?

The results of an experiment to measure the rate of reaction between marble chips and dilute hydrochloric acid are shown in the table.

Time (s)	Volume of gas produced (cm³)
0	0
60	40
120	56
180	62
240	64

This type of graph gives the following information:

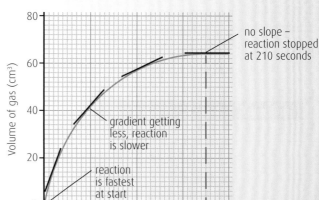

 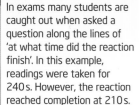
contd

1 The gradient (slope) of the graph indicates the reaction rate. The greater the gradient, the faster the reaction. As the gradient is steepest at the start of the reaction, this indicates the fastest reaction rate. Where the graph is horizontal, the reaction has stopped. As the reaction proceeds, the reaction rate slows down as the reactants gets used up. This is shown by the decreasing gradient.

2 The total volume of gas produced in the reaction was 64 cm³. This is read from the graph at the point where the reaction has stopped.

CALCULATING THE REACTION RATE

DON'T FORGET

You must be able to calculate the average reaction rate from a graph.

EXAMPLE

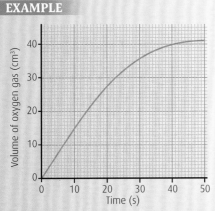

Looking at this graph it can be seen that at time 0 s, the volume of oxygen gas is 0 cm³. After 10 s the volume of oxygen gas has increased to 15 cm³. Therefore, the change in volume over the first 10 s is 15 cm³.

The average rate of the reaction over the first 10 s = $\frac{15 - 0}{10 - 0}$ = 1·5 cm³ s⁻¹

After 10 s to 20 s the volume of oxygen gas has increased from 15 cm³ to 27 cm³.

The average rate of reaction between 10 s and 20 s = $\frac{27 - 15}{20 - 10}$

$$= \frac{12}{10}$$

$$= 1\cdot2 \text{ cm}^3\text{ s}^{-1}$$

The calculations confirm that the reaction is slowing down as it proceeds.

ONLINE

Click the link 'Understanding Reaction Rate Graphs Slideshow' at www.brightredbooks.net/N5Chemistry.

EXAMPLE

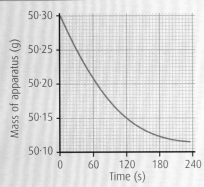

In this example, the reaction rate is calculated exactly as before even though the curve slopes downwards.

From 60 s to 120 s the mass has changed from 50·21 g to 50·15 g.

The average rate of reaction between 60 s and 120 s = $\frac{50\cdot21 - 50\cdot15}{120 - 60}$

$$= \frac{0\cdot06}{60}$$

$$= 0\cdot001 \text{ g s}^{-1}$$

ONLINE TEST

Take the 'Reaction Rates: Graphing the rate of reaction' test at www.brightredbooks.net/N5Chemistry.

REACTION RATES: GRAPHING THE RATE OF REACTION 2

CHANGING THE REACTION CONDITIONS

Using a catalyst or changing the masses, volumes, concentrations or even the particle size of solid reactants involved in a chemical reaction can alter the rate of the reaction.

Consider the reaction between magnesium metal and dilute hydrochloric acid:

$$Mg(s) + 2HCl(aq) \rightarrow MgCl_2(aq) + H_2(g)$$

Changing reaction conditions can be monitored using the aparatus shown. Graphs can then be plotted using the results of the individual experiments.

Looking at the graph it can be deduced that the reaction in experiment 3 was faster than the reaction in experiments 1 and 2 as the slope for experiment 3 is steeper.

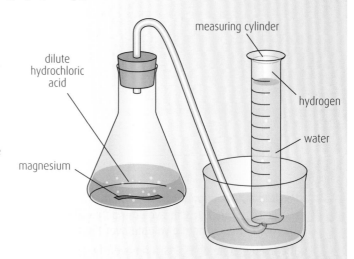

measuring cylinder

dilute hydrochloric acid

hydrogen

water

magnesium

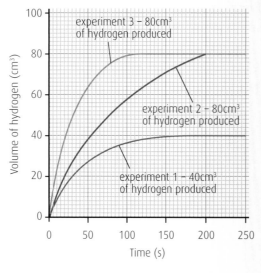

We can also see that the total volume of hydrogen produced in experiments 2 and 3 were twice that of experiment 1.

Why did experiments 2 and 3 produce more gas? The simple answer to this is that a greater quantity of reactants was used.

Experiment 2 could have started with:

- twice the volume of hydrochloric acid.
- twice the concentration of hydrochloric acid.

Experiment 3 could have started with:

- twice the mass of magnesium
- twice the concentration of hydrochloric acid.

contd

DON'T FORGET

A steeper slope means a faster reaction.

DON'T FORGET

The quantity of a product made in a reaction is directly related to the quantities of the reactants used. For example, if a piece of magnesium ribbon and an equal mass of magnesium powder are added to two separate beakers each containing the same volume and concentration of dilute hydrochloric acid, then the quantity of hydrogen produced will be identical.

DON'T FORGET

Adding a catalyst to a reaction or increasing the temperature of a reaction will increase the reaction rate. However, these changes will not increase the final quantity of products made.

ONLINE

Click the 'Introducing Rates of Reaction' link at www.brightredbooks.net/ N5Chemistry.

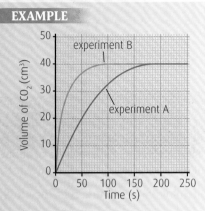

Looking at this graph it can be seen that at time 25 s, experiment A produced 10 cm³ of carbon dioxide while experiment B produced 30 cm³ of carbon dioxide. It is also clear that the slope for experiment B is always steeper than the slope for experiment A. Both these observations indicate that in experiment B the reaction is faster than in experiment A.

The reaction in experiment B could have been faster due to:

- the use of a catalyst
- a higher reaction temperature
- an increased concentration of a reactant
- a decreased particle size of a solid reactant.

ONLINE TEST

Test yourself on reaction rates online in the BrightRED Digital zone.

THINGS TO DO AND THINK ABOUT

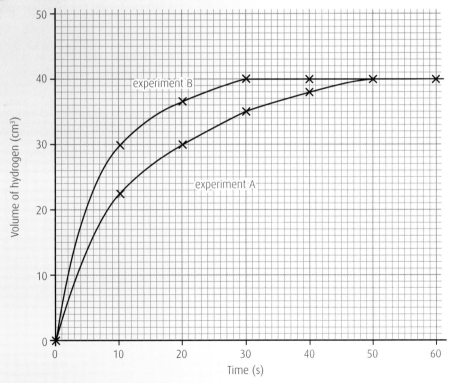

Jill added magnesium to sulfuric acid and measured the volume of hydrogen produced. Her results are displayed on the graph.

(a) Calculate the average rate of reaction for both experiment A and experiment B over the first 10 s.

(b) The same mass of magnesium and the same volume and concentration of sulfuric acid was used in both experiments. How can you tell this from the graph?

(c) Which experiment was carried out at a higher temperature?

ATOMIC STRUCTURE: THE PERIODIC TABLE

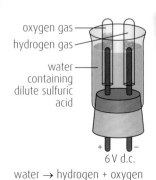

oxygen gas
hydrogen gas
water containing dilute sulfuric acid

6 V d.c.

water → hydrogen + oxygen

ELEMENTS

An **element** is a substance that cannot be broken down into anything simpler by chemical means. As of 2018, there are 118 known elements. Just like the 26 letters of the alphabet make all the words we use, the chemical elements combine to make millions of different chemical compounds. Elements can be thought of as the building blocks of all substances.

When electricity is passed through water containing dilute sulfuric acid, the water breaks down (decomposes) into the elements hydrogen and oxygen. As hydrogen and oxygen are elements, they cannot be broken down any further.

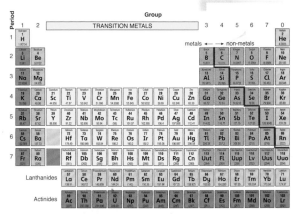

THE MODERN PERIODIC TABLE

All the known elements are arranged on the Periodic Table. Although there were a number of scientists that formulated a 'periodic table', the Russian scientist Dmitri Mendeleev was the first person to have his Periodic Table widely acknowledged by the scientific community.

Each element has a different symbol and a different **atomic number**. The elements are placed in order of their increasing atomic numbers into the Periodic Table, which is arranged in rows (**periods**) and columns (**groups**).

Features of the periodic table

- The majority of the elements are metals.

- Most elements are solids, a few are gases and two (mercury and bromine) are liquids at room temperature.

- The vertical columns are called **groups** and are numbered from 1 to 7, then 0.

- The horizontal rows are called **periods** and are numbered from 1 to 7.

- Elements with similar chemical properties are in the same group.

- The transition metals lie between groups 2 and 3.

- Important 'families' of elements include the alkali metals, the halogens and the noble gases.

 Li Lithium

Na Sodium

K Potassium

Rb Rubidium

Cs Cesium

Fr Francium

GROUP 1 – THE ALKALI METALS

As with all the families of elements, the alkali metals are not identical but they do show similar chemical properties.

The alkali metals are all soft metals which are shiny when freshly cut but lose the shininess when exposed to air, as a layer of oxide forms on the surface:

metal + oxygen → metal oxide

For this reason the metals are stored in liquid paraffin to help prevent contact with the air or with water.

Alkali metals are so-called because they react with water to form alkaline solutions.

> **EXAMPLE**
>
> Sodium reacts vigorously with water to produce the alkali sodium hydroxide. Hydrogen gas is also made in this reaction.
>
> sodium + water → sodium hydroxide + hydrogen
> (an alkali)

GROUP 7 – THE HALOGENS

Like hydrogen, nitrogen and oxygen, the halogen elements are **diatomic molecules** (see 'Formulae and reaction quantities: Chemical formulae' to review diatomic molecules).

Fluorine and chlorine are gases, bromine is a liquid and iodine is a solid. Astatine is a radioactive element that does not occur naturally on Earth.

The halogens and their compounds have many uses. Fluorine compounds are used in toothpaste to help avoid cavities. The most common use of chlorine is in drinking water as it can kill harmful bacteria. Bromine is used in dyes and medicines and iodine is used as an antiseptic.

F
Cl
Br
I
At

ONLINE TEST

Think you know about the Periodic Table? Go online and test yourself at www.brightredbooks.net/N5Chemistry.

GROUP 0 – THE NOBLE GASES

The group 0 elements are all colourless gases. They are extremely unreactive and form almost no known chemical compounds. They do have their uses in lasers, lighting and airships.

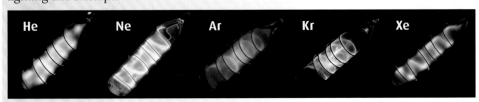

He   Ne   Ar   Kr   Xe

Coloured lights courtesy of the noble gases.

The helium in this airship is much less dense than air.

THINGS TO DO AND THINK ABOUT

1   Are the following statements true or false?
    - Lithium is an alkali metal.
    - Hydrogen is an alkali metal.
    - There are five noble gases.
    - The noble gases are very reactive.
    - Fluorine is the most reactive halogen.
    - The alkali metals are very unreactive.
    - The noble gas argon is used in lightbulbs.
    - The alkali metals are found in group 7.
    - Sulfur is a member of the halogen family.
    - Helium forms lots of compounds with other elements.

2   The elements 114 (Fl) and 116 (Lv) were given their names by the International Union of Pure and Applied Chemistry in May 2012. Try to find out how the names of these elements were decided.

3   Pick an element with an atomic number between 1 and 36 and find out all you can about it. You might include some or all of the following:
    - symbol and atomic number
    - who discovered the element
    - date of discovery
    - main sources
    - uses.
    Create a powerpoint and present your element to the rest of your class.

VIDEO LINK

Learn more by watching 'Noble Gases' at www.brightredbooks.net/N5Chemistry.

DON'T FORGET

The data booklet is an excellent source of information on both elements and the Periodic Table.

ATOMIC STRUCTURE: THE STRUCTURE OF THE ATOM

A cut diamond – one form of the element carbon

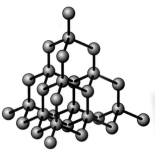

Structure of diamond – all the atoms are the same

THE HISTORY OF THE ATOM

Ancient Greek philosophers originally had the idea that if something was broken down into smaller and smaller pieces eventually it would not be possible to break it down into anything simpler. At the start of the nineteenth century, the English scientist John Dalton (1776–1803) put forward atomic theory to explain his experimental observations. Dalton was the first person to use the word **atom**, which comes from the Greek word atomus, meaning unsplittable.

Dalton suggested that each chemical element had its own unique type of atom, which was different from the atoms of all other elements. He also thought that atoms were like tiny, hard spheres, which could not be broken up.

Between 1897 and 1932, four scientists – Joseph John Thomson, Ernest Rutherford, Niels Bohr and James Chadwick – independently developed Dalton's ideas and through further research found strong experimental evidence to put forward the modern ideas on atomic structure.

THE MODERN ATOMIC MODEL

Atoms are tiny particles that are present in all elements.

The current atomic model indicates that atoms are composed of three even smaller **sub-atomic particles** called **protons**, **neutrons** and **electrons**.

Every atom has an extremely small **nucleus** composed of protons and neutrons. The electrons orbit outside the nucleus.

Chemists now know why atoms react and combine with each other.

An understanding of the properties of sub-atomic particles leads to a greater understanding of how atoms will behave when they are involved in chemical reactions.

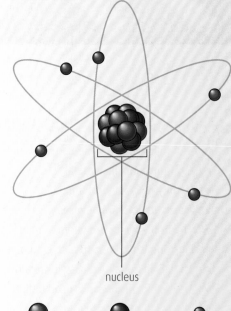

nucleus

proton   neutron   electron

Particle	Position	Relative mass	Charge
Proton	Nucleus	1	+1
Neutron	Nucleus	1	0
Electron	Outside the nucleus	Almost zero	−1

Relative mass

It is clear from the table that protons and neutrons are much heavier than electrons. To balance the mass of one proton or one neutron would require around 1830 electrons. This means that the mass of an atom is concentrated in the nucleus. When the mass of an atom is calculated, the mass of any electrons present is ignored.

DON'T FORGET

Atoms are tiny. One gram of hydrogen atoms would contain more than six hundred thousand million, million, million atoms. As they are so small, ordinary units of mass used in everyday life cannot be used for atoms. Scientists use the atomic mass scale to measure the mass of atoms, protons, neutrons and electrons. The relative mass of a hydrogen atom is 1 atomic mass unit.

contd

Charge

Protons are positively charged particles and electrons are negatively charged particles. As their name suggests, neutrons are neutral and carry no electrical charge.

This means that the nucleus of an atom will be positively charged. All atoms are electrically neutral because the positive charge of the nucleus is equal to the total negative charge of the electrons.

Consider the lithium atom shown.

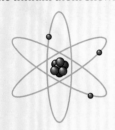

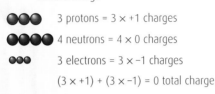

electric charge

3 protons = 3 × +1 charges

4 neutrons = 4 × 0 charges

3 electrons = 3 × −1 charges

(3 × +1) + (3 × −1) = 0 total charge

A lithium atom

ATOMIC NUMBER AND MASS NUMBER

Each element in the Periodic Table has its own **atomic number**. The atomic number of an element is equal to the number of protons in its nucleus. The periodic table arranges the elements in order of increasing atomic number. The first element, with an atomic number of 1, is hydrogen. All hydrogen atoms have one proton in their nucleus.

Every atom is electrically neutral and since the only charged particles present in an atom are protons and electrons, then the number of protons must equal the number of electrons.

Every atom has a **mass number**, which is defined as the total number of protons and neutrons in its nucleus.

Consider the three atoms shown.

Helium

Carbon

Fluorine

Helium: Atomic number is 2 as the atom has two protons. The mass number of this atom is 4 (two protons and two neutrons).

Carbon: Atomic number is 6 as the atom has six protons. The mass number of this atom is 12 (six protons and six neutrons).

Fluorine: Atomic number is 9 as the atom has nine protons. The mass number of this atom is 19 (nine protons and ten neutrons).

THINGS TO DO AND THINK ABOUT

1   Try to find out the contribution the scientists Thomson, Rutherford, Bohr and Chadwick made to our understanding of the structure of the atom.

2   An atom has seven protons, seven electrons and eight neutrons.
    (a) What is the atomic number of this atom?
    (b) What is the mass number of this atom?
    (c) Explain why this atom has no overall charge.

3   An atom of magnesium has a mass number of 25. How many protons, neutrons and electrons does this magnesium atom have?

4   Look again at the diagrams of the atoms above. Draw a similar diagram for an atom that has 11 protons, 11 electrons and 12 neutrons.

DON'T FORGET

Protons are positively charged.
Neutrons are neutral.
Electrons are negatively charged.

VIDEO LINK

Check out the clip 'Basic Atomic Structure' at www.brightredbooks.net/N5Chemistry.

VIDEO LINK

Watch the video 'Structure of an Atom' for more at www.brightredbooks.net/N5Chemistry.

DON'T FORGET

The atomic number defines an element. The number of protons is therefore the only sub-atomic particle that can be used to identify an element.

DON'T FORGET

Never include electrons in the calculation of the mass number of an atom.

DON'T FORGET

The current model of an atom has a positively charged nucleus, containing protons and neutrons, with negatively charged electrons orbiting outside the nucleus. The vast majority of the mass of the atom is in the nucleus, and atoms have no overall electrical charge.

VIDEO LINK

Check out the clip on atomic number and mass number at www.brightredbooks.net/N5Chemistry.

ATOMIC STRUCTURE: NUCLIDE NOTATION OF ATOMS AND IONS

VIDEO LINK

Check out the clip on nuclide symbols at www.brightredbooks.net/N5Chemistry.

VIDEO LINK

Be sure to watch the clip 'Atomic Symbols' at www.brightredbooks.net/N5Chemistry.

DON'T FORGET

Subtracting the atomic number from the mass number will always give the number of neutrons the particle has. This is shown by the equation:
number of neutrons = mass number (A) – atomic number (Z)

DON'T FORGET

In an atom, protons = electrons = atomic number = Z.

DON'T FORGET

Electron arrangements are found in the data booklet.

DON'T FORGET

Elements in the same group of the Periodic Table have similar chemical properties – they react in a similar way. We know that during a reaction the particles involved must collide successfully. Chemists believe that it is the outer energy level electrons that are involved in the collisions and so it is not surprising that elements with the same number of electrons in their outer energy level are chemically similar.

NUCLIDE NOTATION

Chemists use a special symbol known as **nuclide notation** to show the numbers of sub-atomic particles in an atom. Nuclide notation has the form:

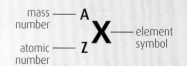

Consider four examples:

$^{7}_{3}\text{Li}$ $^{39}_{19}\text{K}$ $^{16}_{8}\text{O}$ $^{19}_{9}\text{F}$

Particle	Number
proton	3
neutron	4
electron	3

Particle	Number
proton	19
neutron	20
electron	19

Particle	Number
proton	8
neutron	8
electron	8

Particle	Number
proton	9
neutron	10
electron	9

These examples show how the numbers of protons, neutrons and electrons are connected to the nuclide notation. You should be able to determine the nuclide notation given the numbers for each of the sub-atomic particles and be able to deduce the number of sub-atomic particles from a given nuclide notation.

ELECTRON ARRANGEMENT

In an atom, electrons are in constant motion around the nucleus. Chemists believe that the electrons are arranged in the outer parts of the atom in regions of space known as energy levels.

In National 5 chemistry, simplified diagrams of the energy levels called 'target diagrams' can be used to show how the electrons are arranged. The electron arrangements for six of the first 20 elements are shown below.

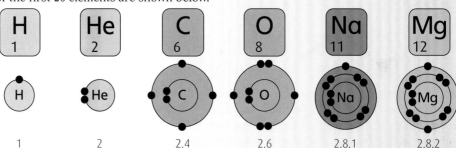

Each electron in an atom must occupy the lowest energy level possible until it is full. A study of the electron arrangements of the first 20 elements reveals a number of important features:

- The first electron energy level can hold a maximum of two electrons, the second and the third shells can both hold a maximum of eight electrons.

- Adding together the individual numbers in an electron arrangement gives the total number of electrons the particle has. For example, sodium has the electron arrangement 2,8,1, which indicates a total of 11 electrons.

- The number of electrons in the outer energy level is the same as the group number of the element. For example, magnesium has the electron arrangement 2,8,2 and like the other group 2 elements there are two electrons in the outer energy level. Note: this is not true for group 0.

IONS

The noble gases are all extremely unreactive elements. It is thought that this is due to the fact that noble gas atoms all have a full or complete outer shell of electrons. A complete outer shell of electrons is a particularly favourable or stable electron arrangement.

During some chemical reactions atoms lose or gain electrons when they collide in order to achieve a stable, full, outer shell of electrons.

When atoms lose or gain electrons they form a new type of particle called an **ion**. Ions can be positively or negatively charged. Ions that possess a **positive charge** have **lost** one or more electrons, while ions that possess a **negative charge** have **gained** one or more electrons. This idea is explained in more detail in the Chemical bonding section of this study guide.

Ions are charged particles. When writing the chemical symbol of an ion, the size and nature of the charge is indicated at the top right-hand corner of the symbol.

2   2,8   2,8,8

 ONLINE TEST

Take the test on nuclide notation of atoms and ions at www.brightredbooks.net/N5Chemistry.

EXAMPLE

$$\text{Na}^+$$
A charge of one positive: This ion has one electron less than a sodium atom.

A sodium ion

$$\text{Mg}^{2+}$$
A charge of two positive: This ion has two electrons less than a magnesium atom.

A magnesium ion

$$\text{F}^-$$
A charge of one negative: This ion has one electron more than a fluorine atom.

A fluoride ion

$$\text{S}^{2-}$$
A charge of two negative: This ion has two electrons more than a sulfur atom.

A sulfide ion

DON'T FORGET

Many students confuse the positive (+) and negative (–) charge signs with the arithmetic symbols plus and minus. As a result of this, when calculating the number of electrons in an ion, they incorrectly add electrons when a positive charge is shown and subtract electrons when a negative charge is shown. Remember that a negative charge will increase the number of electrons and a positive charge will decrease the number of electrons.

Nuclide notation

When ions are produced in a chemical reaction the number of electrons the particle has will change. The number of protons and the number of neutrons will be unaffected by the reaction. The nuclide notation for an ion, like the symbols shown above, will change to reflect the change in the number of electrons.

EXAMPLE

How many protons, neutrons and electrons do the following ions have?

(a) $^{27}_{13}\text{Al}^{3+}$

number of protons = atomic number = 13

number of neutrons = mass number – atomic number
= 27 – 13 = 14

number of electrons = atomic number – charge on the ion = 13 – 3 = 10

This aluminium ion has 13 protons, 14 neutrons and ten electrons.

(b) $^{16}_{8}\text{O}^{2-}$

number of protons = atomic number = 8

number of neutrons = mass number – atomic number
= 16 – 8 = 8

number of electrons = atomic number – charge on the ion = 8 – (–2) = 10

This oxide ion has eight protons, eight neutrons and ten electrons.

 THINGS TO DO AND THINK ABOUT

1 Determine the number of protons, neutrons and electrons in each of the following atoms.

 (a) $^{56}_{26}\text{Fe}$ (b) $^{35}_{17}\text{Cl}$ (c) $^{239}_{94}\text{Pu}$ (d) $^{90}_{36}\text{Kr}$

2 The nuclide notation for an ion of copper is $^{64}_{29}\text{Cu}^{2+}$.

 (a) What is the charge on this copper ion?

 (b) Explain why this copper ion is not electrically neutral.

 (c) This ion was formed from a copper atom. Write the nuclide notation for the copper atom.

 ONLINE

Build your own atom by following the link at www.brightredbooks.net/N5Chemistry!

ATOMIC STRUCTURE: ISOTOPES AND RELATIVE ATOMIC MASS

DON'T FORGET

Isotopes are atoms with the **same atomic number** but **different mass number**. In other words, **isotopes** are atoms with the **same number of protons** but **different numbers of neutrons**.

VIDEO LINK

Check out the video 'What are Isotopes?' at www.brightredbooks.net/N5Chemistry.

DON'T FORGET

All the isotopes of an element are chemically similar. The number of neutrons in an atom has no effect on the chemical properties of the atom.

ONLINE

Take the quiz on atomic structure at www.brightredbooks.net/N5Chemistry.

ISOTOPES

Many textbooks and websites state that an element is a substance that contains one type of atom. This definition of an element is not entirely true. All the atoms of an element have the same atomic number; they contain the same number of protons in their nucleus. For example, an atom that contains one proton is a hydrogen atom and an atom that contains 17 protons is a chlorine atom. All chlorine atoms are similar; they all have 17 protons in their nucleus.

While all atoms of an element will have the same number of protons, the number of neutrons in the nucleus can vary. This means that atoms of an element can have different masses and so have different mass numbers. Atoms of an element that have different mass numbers are called **isotopes**.

All of the chemical elements have isotopes. The element with the highest number of naturally occurring stable isotopes is tin, which has ten isotopes. Those who state that elements have only one type of atom are ignoring the existence of isotopes.

The isotopes of hydrogen

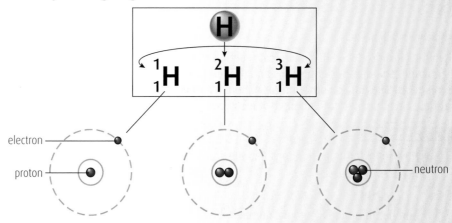

Hydrogen has three naturally occurring isotopes. All three of the hydrogen atoms have one proton (and one electron). The mass numbers of the isotopes are different. This is because the number of neutrons varies from 0 to 1 to 2.

EXAMPLE

The element neon has three stable isotopes. The nuclide notations for two of these isotopes are shown below.

$^{20}_{10}\text{Ne}$ $^{22}_{10}\text{Ne}$

(a) How many neutrons are in each of these isotopes?

number of neutrons = mass number – atomic number

For Ne-20, neutrons = 20 – 10 = 10

For Ne-22, neutrons = 22 – 10 = 12

(b) Write the nuclide notation for the third isotope if it contains 11 neutrons.

All isotopes of neon have the atomic number 10 – they have 10 protons.

mass number = number of protons + number of neutrons

= 10 + 11 = 21

The nuclide notation is therefore $^{21}_{10}\text{Ne}$.

RELATIVE ATOMIC MASS

The masses of atoms can be measured using a complex piece of analytical equipment called a **mass spectrometer**.

The mass spectrometer produces a chart known as a mass spectrum. Analysis of this chart allows chemists to determine not only the mass of each isotope present in an element but also the proportion or abundance of each isotope.

Mass spectrometer

The mass spectrum of chlorine

The mass spectrum indicates that chlorine has two isotopes, one known as chlorine-35 (Cl-35) and the other as chlorine-37 (Cl-37). The spectrum also shows that chlorine-35 atoms are three times more abundant (75%) than chlorine-37 atoms (25%).

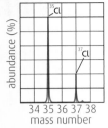

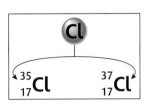

Isotopes of chlorine

The **relative atomic mass** of an element can be calculated from the figures obtained from mass spectrometer data. Relative atomic mass is the average mass of the isotopes present taking into account their relative proportions.

If chlorine contained 100% of chlorine-37, then its relative atomic mass would be 37. If it contained a 50:50 mixture of chlorine-35 and chlorine-37, then its relative atomic mass would be 36.

Using the figures from the mass spectrum gives naturally occurring chlorine a relative atomic mass of 35·5.

The table shows the relative atomic mass for three elements. A more complete list can be found in the data booklet.

Element	Symbol	Relative atomic mass (RAM)	Rounded RAM
Chlorine	Cl	35·5	35·5
Iron	Fe	55·8	56
Magnesium	Mg	24·3	24

DON'T FORGET

The relative atomic mass of an element and the mass number of an atom are not the same thing. Mass number relates to an individual atom and it will always be a whole number. Relative atomic mass is an average calculated from all the isotopes of an element and so it will seldom be a whole number.

As relative atomic masses are calculated averages it is possible to state several facts regarding their value:

- They will rarely be a whole number.

- They will always be **closest to the most abundant isotope** of the element.

- They always lie between the smallest isotopic mass and the largest isotopic mass.

Consider the following data, which gives information about the three isotopes of magnesium.

Studying the data it can be predicted that the relative atomic mass will be between 24 and 26 and that the value will be close to 24 as this is the most common isotope (the isotope with the largest abundance). The actual relative atomic mass of 24·3 confirms the predictions.

Mass number of isotope	Abundance (%)
24	79
25	10
26	11

ONLINE TEST

Revise this topic online and take the 'Isotopes and Relative Atomic Mass' test at www.brightredbooks.net/N5Chemistry.

VIDEO LINK

Check out the video 'How to find the Relative Atomic Mass from Mass Spectral Data' at www.brightredbooks.net/N5Chemistry.

THINGS TO DO AND THINK ABOUT

Copper is made up of two different types of atom:

$^{63}_{29}Cu$ $^{65}_{29}Cu$

(a) What term is used to describe these different types of copper atom?

(b) A sample of copper has a relative atomic mass of 63·5. What is the mass number of the most common type of atom in the sample of copper?

VIDEO LINK

Take the quiz 'Isotopes and Atomic Structure' at www.brightredbooks.net/N5Chemistry.

CHEMICAL BONDING: IONIC BONDING 1

DON'T FORGET

The first electron energy level holds a maximum of two electrons, while the others can hold up to eight electrons. Helium's outer energy level is full even although it contains only two electrons.

DON'T FORGET

The outer electron energy level is sometimes called the **valence energy level** and the electrons that occupy this energy level are called **valence electrons**.

DON'T FORGET

Atoms are electrically neutral as they have the same number of protons and electrons. When there is an imbalance in the number of protons and electrons, the charged particles are known as ions.

DON'T FORGET

Atoms achieve a noble gas electron arrangement when they react, but they **do not** become noble gases. Remember that it is the number of protons which dictates the identity of an element. The sodium ion has the same electron arrangement as a neon atom **but is not a neon atom.** Sodium ions have 11 protons while neon atoms have 10 protons.

ONLINE

Check out the link to 'Ionic Bonding' at www.brightredbooks.net/N5Chemistry!

WHY ATOMS COMBINE

Chemists believe that when atoms react it is the electrons in the outer energy level that are involved. This means that it is the electrons in the outer energy level that are responsible for the chemical behaviour of atoms.

All the noble gas elements have atoms that have a **complete outer energy level** – the energy level contains the maximum permitted number of electrons. The lack of reactivity of the noble gases is due to this fact. A complete outer energy level of electrons is a very **stable electron arrangement** and therefore resistant to change.

helium

Ne
2,8
neon

Ar
2,8,8
argon

The electron arrangements of the first three noble gases

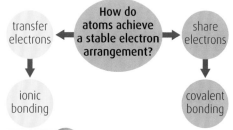

All the other elements in the periodic table have atoms with incomplete outer energy levels and so they will react in order to achieve a stable electron arrangement similar to that of a noble gas.

When atoms react they can achieve a stable electron arrangement in one of two different ways: electron transfer and electron sharing. Electron transfer results in **ionic bonding** and electron sharing results in **covalent bonding**.

IONIC BONDING

Ionic compounds are usually formed when a metal combines with a non-metal(s). During this process, the atoms achieve the stable electron arrangement of a noble gas by the transfer of electrons from the metal atom to the non-metal atom. As a result of the electron transfer, the atoms involved become charged particles called ions.

In general, metal atoms tend to **lose electrons and form positively charged ions**, while non-metals tend to **gain electrons and form negatively charged ions**. The formation of ions through the loss or gain of electrons can be shown in an ion-electron equation.

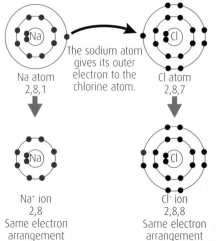

Na atom
2,8,1

The sodium atom gives its outer electron to the chlorine atom.

Cl atom
2,8,7

Na⁺ ion
2,8
Same electron arrangement as neon

Cl⁻ ion
2,8,8
Same electron arrangement as argon

The formation of sodium chloride

When sodium metal reacts with chlorine gas the ionic compound sodium chloride (common salt) is produced.

The ion-electron equation for sodium losing an electron to form a one positive sodium ion, Na^+ is

$$Na \rightarrow Na^+ + e^-$$

2,8,1   2,8

The ion-electron equation for chlorine gaining an electron to form a one negative chloride ion, Cl^-:

$$Cl + e^- \rightarrow Cl^-$$

2,8,7   2,8,8

OTHER IONIC COMPOUNDS

Other ionic compounds are formed in a similar way. However, the charge values on the ions are dependent on the number of electrons an atom needs to lose or gain to achieve a stable electron arrangement.

The groups of the periodic table help to identify the charge on an ion.

During the formation of an ionic compound all the elements in a group will behave in a similar way to the elements shown in the tables. For example, all the metal elements in group 2 will form a two-positive ion when they react and all the non-metal elements in group 7 will form a one-negative ion when they react.

The number of electrons an atom loses or gains to achieve a stable electron arrangement is often called the **valency number** of the element. Valency numbers are the key to writing chemical formulae and are discussed in more detail in the Formulae and reaction quantities section of this study guide.

The formation of calcium oxide

When calcium metal reacts with oxygen the ionic compound calcium oxide is formed. In this case, two electrons are transferred from each calcium atom to each oxygen atom.

Table 1: Metals

Group	1	2	3
Example	Lithium	Magnesium	Aluminium
Electron arrangement of atom	2,1	2,8,2	2,8,3
Number of electrons lost	1	2	3
Ion symbol	Li^+	Mg^{2+}	Al^{3+}
Electron arrangement of ion	2	2,8	2,8
Similar noble gas	Helium	Neon	Neon

Table 2: Non-metals

Group	5	6	7
Example	Phosphorus	Sulfur	Fluorine
Electron arrangement of atom	2,8,5	2,8,6	2,7
Number of electrons gained	3	2	1
Ion symbol	P^{3-}	S^{2-}	F^-
Electron arrangement of ion	2,8,8	2,8,8	2,8
Similar noble gas	Argon	Argon	Neon

$$
\begin{array}{ccccc}
\text{Ca} & + & \text{O} & \rightarrow & \boxed{\begin{array}{ccc} Ca^{2+} & + & O^{2-} \\ 2,8,8 & & 2,8 \\ \text{calcium ion} & & \text{oxide ion} \end{array}} \\
2,8,8,2 & & 2,6 & & \\
\text{calcium atom} & \text{oxygen atom} & & &
\end{array}
$$

2 electrons transferred → calcium oxide

THINGS TO DO AND THINK ABOUT

1 Consider the elements shown in the table.

Potassium	Oxygen	Fluorine
Xenon	Aluminium	Phosphorus

   (a) Which element will not form ions?

   (b) Which element is most likely to form a three-negative ion?

   (c) Which element is most likely to lose three electrons to become stable?

2 When potassium reacts with oxygen it forms the ionic compound potassium oxide.

   (a) What is the electron arrangement of a potassium ion?

   (b) Which noble gas has the same electron arrangement as this potassium ion?

3 When ionic compounds are formed, the reacting atoms achieve a noble gas electron arrangement. Explain why the reacting atoms do not actually become noble gases.

ONLINE TEST

Revise this topic by taking the 'Ionic Bonding' test at www.brightredbooks.net/ N5Chemistry.

CHEMICAL BONDING: IONIC BONDING 2

WHAT IS AN IONIC BOND?

In an ionic compound there will be huge numbers of positive and negative ions. The ions have opposite charges and so they attract each other. This attraction holds the ions together and is known as an ionic bond.

The oppositely charged ions arrange themselves into a giant structure known as an **ionic lattice**. Strong ionic bonds extend throughout the lattice structure.

This diagram shows that each sodium ion is surrounded by chloride ions, and vice versa.

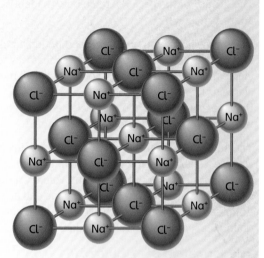

VIDEO LINK

Watch the clip showing the synthesis of NaCl at www.brightredbooks.net/N5Chemistry.

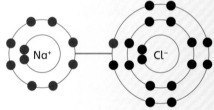

An ionic bond – the electrostatic attraction between oppositely charged ions

PHYSICAL PROPERTIES EXPLAINED

Most things have properties that can be used to identify them. For example, we can identify a person by their face, their eyes, their height and their voice. The more of these properties we can identify, the better we know the person.

Usually ionic compounds are made up of a metal combined with a non-metal, but there are many exceptions to this general rule. By far the best way to identify the bonding and structure in a compound is to know the properties of the compound. Just like identifying a person, the more information we have regarding the properties of a substance the better our decision will be with regard to the type of the bonding the substance has.

DON'T FORGET

Most of the properties of a substance depend on the type of bonding present and the structure adopted by the particles.

PHYSICAL STATE

All ionic compounds are hard solids at room temperature – they have high melting points and boiling points.

The table shows the melting points of three substances.

Name	Bonding	Structure	Melting point (°C)	Boiling point (°C)	State at 25°C
sodium chloride	ionic	lattice	801	1413	solid
calcium oxide	Ionic	lattice	2614	2850	solid
barium chloride	ionic	lattice	963	1560	solid

As previously mentioned the ionic bonds in the lattice are very strong. When melting or boiling ionic compounds, these strong bonds have to be broken, which explains why they have high melting points and boiling points.

ELECTRICAL CONDUCTIVITY

An electric current is a flow of electrically charged particles. Substances that allow a current to flow are called **electrical conductors** and substances that do not allow a current to flow are called non-conductors or **electrical insulators**.

Ionic substances do not conduct when they are solid but do conduct if they are molten (melted) or dissolved in water.

Using this apparatus the bulb will light if the substance tested is a conductor.

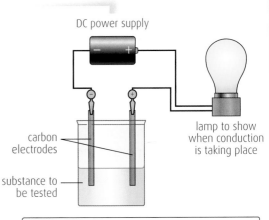

DC power supply

carbon electrodes

substance to be tested

lamp to show when conduction is taking place

Explaining ionic conductivity

When solid, the ions in the lattice are held tightly together and they are not free to move. When the ionic substance is dissolved in water, the lattice breaks up and the ions become free to move. The positive ions will move towards the negative electrode while the negative ions will move towards the positive electrode. This explains why ionic substances do not conduct when solid but do conduct when dissolved. The ions will also be free to move if the substance is molten, as the ionic lattice is broken up by heat.

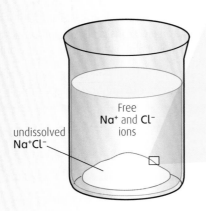

undissolved Na^+Cl^-

Free Na^+ and Cl^- ions

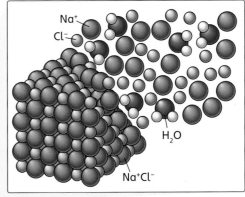

Na^+
Cl^-
H_2O
Na^+Cl^-

SUMMARY

When an ionic compound is formed:

- metal atoms lose electrons and form positive ions

- non-metal atoms gain electrons and form negative ions

- the number of electrons lost or gained is known as the valency number

- the size of the charge an ion has is equal to its valency number

- the ions have a stable electron arrangement identical to the nearest noble gas

- the oppositely charged ions form a lattice structure held together by strong ionic bonds.

In general, ionic compounds:

- have high melting points

- only conduct electricity when molten or dissolved in water.

 # THINGS TO DO AND THINK ABOUT

1   The compound tin iodide is made of a metal and a non-metal. It is an orange crystalline solid at room temperature with a melting point of 144°C. Tin iodide does not conduct electricity when solid or when molten. Identify the type of bonding and structure in tin iodide. Give the reasons for your answer.

2   In general, ionic substances have high melting points. Try to find out the factors that affect the strength of ionic bonds.

CHEMICAL BONDING: COVALENT BONDING

Covalent bonds form when non-metal atoms react with each other. The atoms of non-metals need electrons to achieve a noble gas electron arrangement. In ionic bonding this is achieved by electron transfer from a metal atom. In covalent bonding this cannot happen since both atoms need electrons and so the atoms share electrons. Covalent bonding is the result of the atoms sharing a pair (or pairs) of electrons.

DON'T FORGET

The number of covalent bonds (valency number) a non-metal atom forms can be found using the relationship: valency = 8 – group number. For example, oxygen atoms can make 8 – 6 = 2 covalent bonds. Hydrogen is an exception to this rule. All hydrogen atoms make one covalent bond.

UNPAIRED ELECTRONS

Only outer shell electrons are involved in chemical bonding. In covalent bonding it is important to understand the concept of **unpaired electrons**. The number of unpaired electrons a non-metal atom has will determine the number of covalent bonds the atom can make. The number of unpaired electrons is also equal to the valency number of the atom. The groups of the periodic table help to identify how many unpaired electrons an atom has.

The table shows examples for each non-metal periodic table group. All the atoms of a group will have the same number of unpaired electrons.

Group	1	4	5	6	7	0
Outer electron arrangement	H	C	N	O	F	Ne
Number of unpaired electrons	1	4	3	2	1	0
Number of covalent bonds made	1	4	3	2	1	0

DON'T FORGET

A molecule is a group of atoms joined together by covalent bonds.

DON'T FORGET

The SQA will also accept these drawings as all dots or all crosses.

DON'T FORGET

While the single line represents a covalent bond it will not merit a mark in a question that asks you to clearly show the arrangement of outer electrons in a molecule.

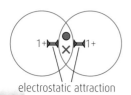

electrostatic attraction

HOW COVALENT BONDS FORM

When covalent bonds form, non-metal atoms join together and form molecules or networks when the unpaired electrons the atoms have are shared.

Chemists use 'dot-and-cross' diagrams to show how covalent bonds form. The shared electron from one atom is shown as a dot, while the shared electron from the other atom is shown as a cross.

EXAMPLE

Hydrogen: H_2

   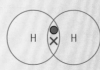

Each hydrogen atom has one unpaired electron in its outer shell. By sharing these two electrons both atoms have a full, stable outer shell like helium.

The hydrogen molecule can also be represented as H–H or H_2. The single line shown represents the covalent bond between the atoms.

What holds the atoms together?

In a covalent bond the two atoms are held together by the common attraction of the two positively charged nuclei for the shared pair of negatively charged electrons. In the hydrogen molecule shown, the electrostatic attraction of protons for electrons 'glues' the atoms together.

contd

EXAMPLE

Chlorine: Cl_2

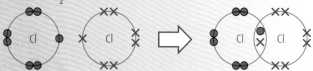

Outer electrons of chlorine

Each chlorine atom has seven outer electrons, only one of which is unpaired. Each atom shares the unpaired electron and so both atoms have a full, stable outer shell like argon.

EXAMPLE

Oxygen: O_2

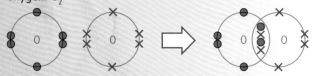

Outer electrons of oxygen

Oxygen atoms have two unpaired electrons and so can form two covalent bonds. In an oxygen molecule, each atom has a share of eight electrons and so has a full, stable outer shell. Oxygen does this by forming a double covalent bond between its atoms.

The oxygen molecule can also be represented as O=O or O_2.

EXAMPLE

Nitrogen hydride (ammonia): NH_3

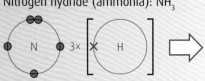

In an ammonia molecule, the nitrogen atom makes three single covalent bonds to three separate hydrogen atoms.

EXAMPLE

Methane: CH_4

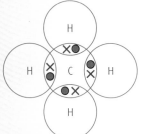

One carbon atom has joined to four hydrogen atoms. The valency number of carbon is four. Atoms of carbon will always make four covalent bonds as they have four unpaired electrons.

DON'T FORGET

A single covalent bond is one shared pair of electrons. A double covalent bond is two shared pairs of electrons.

DON'T FORGET

In a dot-and-cross diagram hydrogen will always have a maximum of two electrons. The atoms of all other elements must have a total of eight electrons (a stable octet). The number of shared pairs (covalent bonds) will be the same as the valency number of the atom.

ONLINE

Check out more on this by clicking on 'Covalent bonding' at www.brightredbooks.net/N5Chemistry.

VIDEO LINK

Watch 'Ionic and Covalent Bonding Animation' at www.brightredbooks.net/N5Chemistry.

ONLINE TEST

Revise this topic by taking the 'Covalent Bonding' test at www.brightredbooks.net/N5Chemistry.

THINGS TO DO AND THINK ABOUT

1 What holds atoms together in a covalent bond?

2 Look again at the dot-and-cross diagrams shown and draw similar diagrams for water (H_2O) and nitrogen fluoride (NF_3).

CHEMICAL BONDING: COVALENT BONDING

The vast majority of covalent substances adopt a structure called a discrete covalent molecular structure, which is a collection of small, separate molecules. A few covalent substances form structures known as covalent networks in which the atoms are joined by covalent bonds that go on and on in three dimensions throughout the whole structure.

PHYSICAL STATE

Covalent substances can be solid (usually softer than ionic solids), liquid or gas at room temperature – they have a wide range of melting points and boiling points.

Name	Bonding	Structure	Melting point (°C)	Boiling point (°C)	State at 25°C
carbon (diamond)	covalent	network	–	3500	solid
silicon dioxide	covalent	network	1700	2300	solid
carbon dioxide	covalent	molecular	–57	–78	gas
water	covalent	molecular	0	100	liquid
sulfur	covalent	molecular	113	445	solid

ELECTRICAL CONDUCTIVITY

Solubility

Covalent substances which do not dissolve in water may dissolve in other solvents such as hexane or carbon tetrachloride.

To conduct electricity a substance must allow a flow of charged particles. As covalent compounds have no free electrons or ions they do not conduct electricity in any state.

COVALENT STRUCTURES AND MOLECULAR SHAPES

Covalent network

All covalent networks have very high melting points and boiling points and like ionic substances they are hard, crystalline solids. Very strong covalent bonds extend throughout the whole network. Melting or boiling this network requires a lot of energy to break the covalent bonds and explains why network substances have high melting points and boiling points.

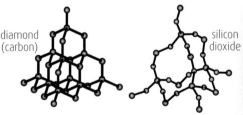

diamond (carbon)

silicon dioxide

Diamond, one form of the element carbon (C), and silicon dioxide (SiO_2 – sand) are two examples of covalent networks.

In both these structures the atoms form one large network of atoms, not a collection of a large number of individual molecules.

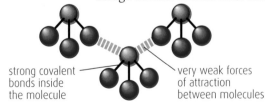

strong covalent bonds inside the molecule

very weak forces of attraction between molecules

Discrete covalent molecular

Most covalent substances have a discrete molecular structure. In this type of substance there are lots of individual (discrete) molecules. The diagram shows

contd

that there are two different types of bonding force in a molecular structure. The bonds inside the molecules are strong covalent bonds. The bonds between the molecules are very weak.

When this type of substance is supplied with energy the individual molecules move further apart.

In melting and boiling only the bonds between the molecules are broken, not the covalent bonds inside the molecules. Since the bonds between the molecules are weak, very little energy is needed to break them and this explains why covalent molecular substances have low melting and boiling points.

Imagine the molecules in a covalent liquid as golf balls in a basket. Shaking the basket would give the liquid more energy and cause it to boil. The individual golf ball molecules would not break up but they would leave the basket. A similar thing happens when water boils. The water molecules do not break apart but they do escape their container.

Discrete golf ball molecules

 DON'T FORGET

The element carbon in the form of the covalent network graphite is an electrical conductor. Graphite is also relatively unreactive and for these reasons it is often used as the electrodes in an electric circuit.

DISCRETE COVALENT MOLECULAR STRUCTURES

Water (H_2O) is a **discrete covalent molecular** compound. There are many discrete H_2O molecules in a sample of water. The molecules are separate entities, although collectively they form part of the whole substance.

Dot-and-cross diagrams of covalent molecules do not show the true shape of the molecule. The shape of a molecule depends on the number of atoms and covalent bonds in the molecule. The shapes of some molecular substances are shown in the table below.

A dotted line shows a bond going behind the paper. A wedge shows a bond coming out of the paper.

Compound	Formula	Shape	Description of shape
hydrogen fluoride	HF	H—F	linear
water	H_2O		angular
ammonia	NH_3		trigonal pyramidal
methane	CH_4		tetrahedral

 DON'T FORGET

To determine the bonding and structure in a substance it is much more reliable to base any conclusion on the properties of the substance. For example, it would be unwise to label a substance as ionic based solely on the fact it contained a metal and a non-metal.

VIDEO LINK

Check out the clip 'Types of Bonds' at www.brightredbooks.net/N5Chemistry.

SUMMARY

- In covalent bonding, valence electrons are shared between the atoms in the bond.

- Covalent substances can have a network or discrete molecular structure.

- Covalent network substances have very high melting points and boiling points.

- Discrete covalent molecular substances have a wide range of melting points and boiling points which are much lower the covalent networks.

- With the exception of carbon (graphite) covalent substances do not conduct electricity in any state.

 ONLINE TEST

Test yourself on this topic at www.brightredbooks.net/N5Chemistry.

 THINGS TO DO AND THINK ABOUT

Covalent substances exist as discrete molecules or networks. Using an example of each type of structure describe the difference between them.

FORMULAE AND REACTION QUANTITIES: CHEMICAL FORMULAE

FORMULAE OF ELEMENTS

The chemical formula of the majority of elements is identical to its symbol. For example, the formula of magnesium is Mg and the formula of iron is Fe.

Certain elements exist as diatomic molecules. In diatomic molecules, the atoms are joined together in pairs – they are **two-atom molecules**.

The chemical formula of a diatomic element is therefore written as X_2, where X is the symbol of the element.

hydrogen (H_2)   nitrogen (N_2)   oxygen (O_2)   fluorine (F_2)   chlorine (Cl_2)   bromine (Br_2)   iodine (I_2)

The seven diatomic elements and their formulae

FORMULAE OF COMPOUNDS

Water is a chemical compound formed from the elements hydrogen and oxygen. The chemical name for water is hydrogen oxide. The chemical formula of water is H_2O. This chemical formula can be represented in a variety of ways.

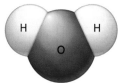
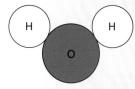
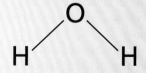

The structures show that one water molecule is made up of **two hydrogen atoms** and **one oxygen atom** – H_2O.

All covalent molecular formulae do this; they show the number and type of each of the atoms that are chemically joined in a compound. The examples shown on the right illustrate this point.

carbon dioxide	CO_2	one carbon atom, two oxygen atoms	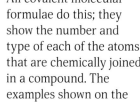
methane	CH_4	one carbon atom, four hydrogen atoms	
ammonia	NH_3	one nitrogen atom, three hydrogen atoms	
ethanol	C_2H_6O	Two carbon atoms, six hydrogen atoms, one oxygen atom	

FORMULAE AND VALENCY

When atoms combine, new chemical bonds are formed. Diagrams of molecules often show the bonds between atoms as solid lines connecting the symbols of the elements.

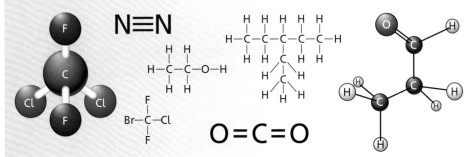

The atoms of most elements usually form the same number of bonds when they make chemical compounds. Looking at the structures shown it is evident that hydrogen, chlorine, fluorine and bromine form one bond, oxygen forms two bonds, nitrogen forms three bonds and carbon forms four bonds.

The **valency** or **valence number** of an atom is defined as the number of bonds an atom can make. Valency is often called the combining power of an atom.

Valency and the periodic table

The valency number of an element can be found from the groups in the periodic table.

Group number	1	2	3	4	5	6	7	0
Valency number	1	2	3	4	3	2	1	0

Elements with variable valencies

Some elements, particularly the transition metals, can have more than one valency number. This is because they can form ions with different charges. The names of transition metal compounds usually indicate the valency of the metal with a **Roman numeral** after the element's name.

EXAMPLE

Copper(II) oxide – copper has a valency of two in this compound.

Iron(III) hydroxide – iron has a valency of three in this compound.

THINGS TO DO AND THINK ABOUT

1. What are the valency numbers of the following elements?
   (a) Silicon   (c) Aluminium   (e) Argon
   (b) Potassium   (d) Carbon

2. Caffeine is a stimulant found in tea and coffee. The molecular formula of caffeine is $C_8H_{10}N_4O_2$.
   (a) How many different elements are in caffeine?
   (b) How many atoms make up one caffeine molecule?

3. Hydrogen cyanide is a very toxic gas. A possible structure of a hydrogen cyanide molecule is shown below.

   $$H-C\equiv N$$

   (a) What is the formula of hydrogen cyanide?
   (b) Explain why chemistry students would expect hydrogen cyanide to contain only two elements.

 ONLINE TEST

Try taking the 'Chemical formulae and reaction quantities' test at www.brightredbooks.net/N5Chemistry.

 DON'T FORGET

In an ionic compound the valency will equal the charge on the ion. For example, Na^+ and K^+ have a valency of one. Mg^{2+} and Fe^{2+} have a valency of two. Cl^- and OH^- have a valency of one. O^{2-} and SO_4^{2-} have a valency of two.

 DON'T FORGET

Roman numerals: one, I; two, II; three, III; four, IV; five, V; six, VI

 ONLINE

Follow the 'Chemical Formulae' link at www.brightredbooks.net/N5Chemistry.

FORMULAE AND REACTION QUANTITIES: WRITING FORMULAE OF COMPOUNDS 1

The chemical formula of a compound can be written either by using only the name of the substance or by using the name of the substance and a set of valency rules.

USING ONLY THE NAME

You can use this method to write a chemical formula when the name of the compound contains prefixes like mono-, di-, tri-, tetra-, etc. The table shows the meaning of several commonly used prefixes:

Prefix	mono	di	tri	tetra	penta	hexa
Meaning	1	2	3	4	5	6

The formula of dinitrogen monoxide can be written straight from its name. The name indicates that the compound contains two nitrogen atoms and one oxygen atom and so its formula is N_2O.

DON'T FORGET

The number 1 is usually omitted in a chemical formula.

Some more examples

phosphorus pentachloride – PCl_5

carbon dioxide – CO_2

boron trihydride – BH_3

silicon tetrafluoride – SiF_4

nitogen dioxide – NO_2

DON'T FORGET

Consider the aluminium chloride example. At the select symbols stage of the process, pupils often incorrectly write Cl_2 because chlorine is diatomic. The formula of chlorine gas as an element is indeed Cl_2. However, the valency of chlorine refers to a chlorine atom or a chloride ion and not a chlorine molecule. This is why the symbol Cl is used in this example. This holds true for all the other diatomic elements.

USING VALENCY RULES 1

Valency rules are a step-by-step process that enables the formulae of many compounds to be written. The first two examples show how to write the correct formulae of lithium oxide and aluminium chloride.

EXAMPLE 1 and 2

	Lithium oxide		Aluminium chloride	
Select symbols	Li	O	Al	Cl
Cross valency numbers	1	2	3	1
Write the formula	Li_2O		$AlCl_3$	

The two examples which follow demonstrate the idea of "cancelling down" valency numbers:

DON'T FORGET

Valency numbers are always cancelled down to give the simplest whole numbers in the formula. For example, valencies of 4 and 2 cancel to 2 and 1, and valencies of 3 and 3 cancel to 1 and 1.

EXAMPLE 3 and 4

	Magnesium oxide		Lead sulfide	
Select symbols	Mg	O	Pb	S
Cross valency numbers	2	2	4	2
Cancel down valencies	Mg₂	O₂	Pb₂	S₄
Write the formula	MgO		PbS_2	

ONLINE

Take the quiz 'Chemical Formulae for Compounds' at www.brightredbooks.net/N5Chemistry.

contd

The two examples which follow consider the formulae of compounds when the name of the compound contains a Roman numeral.

EXAMPLE 5 and 6

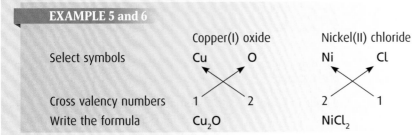

	Copper(I) oxide	Nickel(II) chloride
Select symbols	Cu O	Ni Cl
Cross valency numbers	1 2	2 1
Write the formula	Cu_2O	$NiCl_2$

Many students write incorrect formulae when the name of the compound contains a Roman numeral.

Look again at example 6.

A common incorrect answer for the formula of nickel(II) chloride is Ni_2Cl. You can see that this formula arises because the student has failed to cross over the valency numbers and simply written the formula in the same order as suggested by the name of the compound.

Two final examples

Study these carefully to make sure you understand how to apply the basic valency rules:

EXAMPLE 7 and 8

	Iron(III) sulfide	Aluminium nitride
Select symbols	Fe S	Al N
Cross valency numbers	3 2	3 3
Cancel down valencies	Not possible	Al_3 N_3
Write the formula	Fe_2S_3	AlN

THINGS TO DO AND THINK ABOUT

1   Write the formula for the following compounds.

   (a) Dintrogen tetroxide

   (b) Sulfur hexafluoride

   (c) Manganese dioxide

2   Write the formulae for sodium oxide, calcium chloride and aluminium sulfide

3   Write the formulae for vanadium(III) fluoride, copper(II) oxide and zinc(II) chloride

FORMULAE AND REACTION QUANTITIES: WRITING FORMULAE OF COMPOUNDS 2

USING VALENCY RULES 2

If the name of a compound ends in '...ate' or '...ite' it contains what is known as a complex (group) ion. Examples of these are carbonate, CO_3^{2-}, sulfate, SO_4^{2-} and nitrate, NO_3^-. Ammonium, NH_4^+ and hydroxide, OH^- are also common complex ions. The overall charge on these ions is equal to the valency number of the ion. The formulae of these group ions can be found in the data booklet.

Consider the sulfate ion and the ammonium ion.

The examples below show how to write the correct formulae of ammonium sulfate and zinc(II) nitrate.

> **EXAMPLE 1 and 2**

	Ammonium sulfate		Zinc(II) nitrate	
Select symbols	NH_4	SO_4	Zn	NO_3
Add brackets to complex ions	(NH_4)	(SO_4)	Zn	(NO_3)
Cross valency numbers	(NH_4)	(SO_4)	Zn	(NO_3)
	1	2	2	1
Write the formula	$(NH_4)_2SO_4$		$Zn(NO_3)_2$	

Using brackets correctly in these formulae is essential. Writing the formula of zinc(II) nitrate as $ZnNO_{32}$ is incorrect, as is $Zn(NO)_{32}$. The correct formula shows one zinc ion and two nitrate groups. However, as in the case of ammonium sulfate, if there is only one unit of the complex ion, the bracket is not required.

> **EXAMPLE 3 and 4**

	Magnesium carbonate		Calcium hydroxide	
Select symbols	Mg	CO_3	Ca	OH
Add brackets to complex ions	Mg	(CO_3)	Ca	(OH)
Cross valency numbers	Mg	(CO_3)	Ca	(OH)
	2	2	2	1
Cancel down valencies	Mg	(CO_3)	not possible	
Write the formula	$MgCO_3$		$Ca(OH)_2$	

USING VALENCY RULES 3

Sometimes you will be asked to write the ionic formula for a compound. In such a formula the charges on the ions must be shown.

In an ionic formula the size of the charge on an ion is equal to the valency number of the element or complex ion. If you have forgotten about ions, now would be a good time to look back at page 17.

The charges on ions of the main group elements is summarised as follows:

Group number	1	2	3	5	6	7
Charge on ion	1+	2+	3+	3−	2−	1−

The charges on complex ions are found in the data booklet.

The same basic valency rules are followed when writing ionic formulae. Study the four examples shown below:

EXAMPLE 5 and 6

	Magnesium chloride	Copper oxide
Select symbols including charges and add brackets to both ions	(Mg^{2+}) (Cl^{-})	(Cu^{2+}) (O^{2-})
Cross valency	2 1	2 2
Cancel down valencies	not possible	$(Cu^{2+})_2$ $(O^{2-})_2$
Write the formula	$Mg^{2+}(Cl^{-})_2$	$Cu^{2+}O^{2-}$

EXAMPLE 7 and 8

	Iron(III) sulfate	Calcium chromate
Select symbols including charges and add brackets to both ions	(Fe^{3+}) (SO_4^{2-})	(Ca^{2+}) (CrO_4^{2-})
Cross valency	3 2	2 2
Cancel down valencies	not possible	$(Ca^{2+})_2$ $(CrO_4^{2-})_2$
Write the formula	$(Fe^{3+})_2 (SO_4^{2-})_3$	$Ca^{2+}CrO_4^{2-}$

 THINGS TO DO AND THINK ABOUT

1 Write the ionic formulae for calcium carbonate, sodium dichromate and copper(II) bromide.

2 Writing the ionic formulae for ionic compounds is a daunting task for many students. Write a selection of do's and don'ts for your classmates which would help them overcome the problems associated with writing ionic formulae.

DON'T FORGET

The brackets are omitted in the final formula if the number 1 appears outside the brackets. For example in magnesium chloride the formula is written as $Mg^{2+}(Cl^{-})_2$ not as $(Mg^{2+})_1(Cl^{-})_2$ or $(Mg^{2+})(Cl^{-})_2$

DON'T FORGET

Only valency numbers are cancelled. Never cancel down the charges on the ions. The ionic formula of copper(II) oxide is $Cu^{2+}O^{2-}$ not Cu^+O^-

DON'T FORGET

Never remove a bracket from any ion if the number outside the bracket is greater than 1.

ONLINE TEST

Take the 'Writing formulae of compounds' test online at www.brightredbooks.net/N5Chemistry.

FORMULAE AND REACTION QUANTITIES: THE MOLE 1 – GRAM FORMULA MASS

DON'T FORGET

The formula mass is found by adding together all the relative formula masses of the atoms or ions present in the substance.

DON'T FORGET

As the formula mass is found from relative atomic masses it has no units. For example, we would state the formula mass of water (H_2O) as 18.

DON'T FORGET

To help be consistent with formula mass calculations you should give each element its own set of brackets at Step 1.

ONLINE TEST

Try taking the 'Gram formula mass' test at www.brightredbooks.net/N5Chemistry.

DON'T FORGET

Don't forget that the formula mass is the mass of all the atoms in the formula. When dealing with diatomic elements, candidates in exams often forget to **double** the appropriate relative atomic mass. For example, the formula mass of chlorine (Cl_2) is (**2** × 35·5) = 71. Similarly, the formula mass of oxygen gas (O_2) is 32 and not 16.

VIDEO LINK

Check out the clip about formula mass for more at www.brightredbooks.net/N5Chemistry.

FORMULA MASS

As you learned in the section on atomic theory, all atoms have mass. On the atomic mass scale hydrogen atoms have a relative atomic mass of 1. Oxygen atoms have a relative atomic mass of 16. The **formula mass** of a substance can be found from its formula if we know the relative atomic masses of all the atoms in the substance.

There is a table of relative atomic masses in your data booklet.

EXAMPLE

Calculate the formula mass of calcium bromide, $CaBr_2$.

Step 1 – Identify the number of atoms $\quad (1 \times Ca) + (2 \times Br)$

one calcium in formula	RAM of calcium	two bromines in formula	RAM of bromine

Step 2 – Change the symbols to the relative atomic mass $\quad (1 \times 40) + (2 \times 80)$

Step 3 – Do the arithmetic $\quad 40 + 160$

Step 4 – Final answer $\quad 200$

EXAMPLE

Calculate the formula mass of sodium oxide, Na_2O.

two sodiums in formula	RAM of sodium	one oxygen in formula	RAM of oxygen

Formula mass = $(2 \times 23) + (1 \times 16)$
$\quad\quad\quad\quad = 46 + 16$
$\quad\quad\quad\quad = 62$

EXAMPLE

Calculate the formula mass of potassium carbonate, K_2CO_3.

two potassiums in formula	RAM of potassium	three oxygens in formula	RAM of oxygen

Formula mass = $(2 \times 39) + (1 \times 12) + (3 \times 16)$
$\quad\quad\quad\quad = 78 + 12 + 48$
$\quad\quad\quad\quad = 138$

one carbon in formula	RAM of carbon

FORMULA MASS: DEALING WITH BRACKETS

Many chemical formulae have brackets to show complex ions. For example, $Mg(NO_3)_2$, $Ca(OH)_2$ and $(NH_4)_2SO_4$.

You should always 'multiply out' the brackets to calculate the formula mass. The number of atoms inside the bracket is multiplied by the subscript number which follows the bracket.

EXAMPLE

Calculate the formula mass of magnesium nitrate, $Mg(NO_3)_2$.

In this example is essential to realise that NO_3 indicates **one** nitrogen atom and **three** oxygen atoms. However, in magnesium nitrate there are **two** nitrate groups, shown by $(NO_3)_2$. This indicates **two** nitrogen atoms and **six** oxygen atoms.

Formula mass of $Mg(NO_3)_2$ = $(1 \times Mg) + (2 \times N) + (6 \times O)$
$\quad\quad\quad\quad\quad\quad\quad\quad = (1 \times 24·5) + (2 \times 14) + (6 \times 16)$
$\quad\quad\quad\quad\quad\quad\quad\quad = 24·5 + 28 + 96 = 148·5$

GRAM FORMULA MASS AND THE MOLE

Chemists use the term **mole** to mean an amount of a substance. One mole of any chemical is the formula mass expressed in grams and it is known as the **gram formula mass (GFM)**.

The steps necessary to calculate the gram formula mass are identical to those for formula mass. The unit, g, for grams is now included in the final answer.

EXAMPLE

Calculate the mass of one mole of copper(II) sulfate, $CuSO_4$.

GFM $CuSO_4 = (1 \times Cu) + (1 \times S) + (4 \times O)$
$= (1 \times 63 \cdot 5) + (1 \times 32) + (4 \times 16) = 63 \cdot 5 + 32 + 64 = 159 \cdot 5\,g$

One mole of copper(II) sulfate has a mass of $159 \cdot 5\,g$.

USING THE MOLE

Given the mass of a substance you must be able to calculate the number of moles of the substance and vice versa. The relationship between the number of moles (mol) of a substance and its mass is:

number of moles = mass/GFM

This relationship can also be shown using a formula triangle:

Mass to moles

EXAMPLE

Calculate the number of moles of water in 36 g of water (H_2O).

Step 1 – Use the triangle to write down the correct relationship

number of moles (n) = mass/GFM

Step 2 – Calculate the GFM of H_2O

GFM $= (2 \times 1) + (1 \times 16) = 2 + 16 = 18\,g$

Step 3 – Substitute figures into the relationship

$n = \frac{36}{18} = 2\,mol$

Moles to mass

EXAMPLE

Calculate the mass of 0·8 moles of iron(III) sulfide (Fe_2S_3).

Step 1 – Use the triangle to write down the correct relationship

mass $= n \times$ GFM

Step 2 – Calculate the GFM of Fe_2S_3

GFM $= (2 \times 56) + (3 \times 32) = 112 + 96 = 208\,g$

Step 3 – Substitute values into the relationship

mass $= 0 \cdot 8 \times 208 = 166 \cdot 4\,g$

DON'T FORGET

Always check your answers. Consider the water example. If **one mole** of water has a mass of 18 g, it makes sense that the mass of **two moles** is 36 g – twice as much. An answer of 9 g makes no sense and was probably achieved by using the relationship 'upside-down' – GFM/mass.

THINGS TO DO AND THINK ABOUT

1   Calculate the number of moles in:

   (a) 14 g of KOH   (b) 5 g of $CaCO_3$ (c) 11 g of CO_2.

2   Calculate the mass of:

   (a) 2 mol of KBr   (b) 0·1 mol of $C_6H_{12}O_6$ (c) 0·01 mol of Na_2SO_4.

ONLINE

Take the quiz on the basics of moles, mass and formula mass at www.brightredbooks.net/N5Chemistry.

FORMULAE AND REACTION QUANTITIES: THE MOLE 2 – CONCENTRATION

Standard solutions in volumetric flasks

SOLUTIONS AND CONCENTRATION

Many chemical reactions take place in **solution**. Adding a **soluble** chemical, the **solute**, to a **solvent** will produce a solution. When water is the solvent, an **aqueous** solution is produced.

The table shows some terms associated with chemical solutions:

Term	Definition
Soluble	The ability to dissolve in a solvent (e.g. water)
Insoluble	The inability to dissolve in a solvent (e.g. water)
Solute	The chemical that dissolves in a solvent
Solvent	The chemical that dissolves the solute
Solution	The mixture formed when a solute dissolves
Solubility	A measure of how well a solute will dissolve in a particular solvent
Saturated	A solution into which as much solute as possible has dissolved

The **concentration** of a solution is a measure of the quantity of solute dissolved in a given volume of solution. Adding more solute to a solution will **increase** the concentration. Adding more solvent to a solution will **decrease** the concentration.

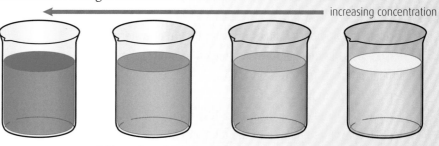

increasing concentration

A blue-coloured solution with different concentrations

Knowing the concentration of a solution to be used in a chemical reaction is very important. The concentration of the solution can affect the speed of the reaction and how much product is made.

A **standard solution** is a solution of known concentration. Standard solutions are usually made up in volumetric flasks.

CONCENTRATION CALCULATIONS

The concentration of a solution is a ratio between the quantity of dissolved solute and the volume of the solution. Chemists express this ratio as a number of moles of solute per litre of solution (mol l⁻¹):

concentration (C) = number of moles of solute (n)/volume of solution in litres (V)

This relationship can be represented by a formula triangle:

A 2 mol l⁻¹ solution is twice as concentrated as a 1 mol l⁻¹ solution and ten times as concentrated as a 0·2 mol l⁻¹ solution.

In examples 1 to 3 on the next page you will be given two of the three quantities in the triangle and asked to determine the third one.

Concentration triangle

contd

EXAMPLE 1, 2 and 3

1. Calculate the **concentration**, in $mol\,l^{-1}$, of $200\,cm^3$ of potassium nitrate solution containing $0.1\,mol$ of potassium nitrate?

   Step 1 – Use the triangle to write down the correct relationship

   $$C = \frac{n}{V}$$

   Step 2 – Substitute values into the relationship

   $C = \frac{0.1}{0.2} = 0.5\,mol\,l^{-1}$ $\boxed{\frac{200}{1000} = 0.2;\ \text{this is done to change } cm^3 \text{ to litres}}$

2. How many **moles** of solute are contained in 3 litres of a $2\,mol\,l^{-1}$ solution of sodium hydroxide?

   Step 1 – Use the triangle to write down the correct relationship

   $n = C \times V$

   Step 2 – Substitute values into the relationship

   $n = 2 \times 3 = 6\,mol$ $\boxed{\text{3 as the question gives the volume in litres}}$

3. What **volume** of $1.5\,mol\,l^{-1}$ glucose solution contains 0.30 moles of glucose?

   Step 1 – Use the triangle to write down the correct relationship

   $V = \frac{n}{C}$

   Step 2 – Substitute values into the relationship

   $V = \frac{0.30}{1.5} = 0.2$ litres $\boxed{\text{As the concentration is given in } mol\,l^{-1}, \text{ the volume calculated is in litres}}$

In these more complex examples the questions will refer to the **mass of the solute**.

EXAMPLE 4 and 5

4. Calculate the concentration of a solution that contains $1.12\,g$ potassium hydroxide (**KOH**) in $250\,cm^3$ of solution.

   Step 1 – Use the 'mass' triangle to find the number of moles (n) of **KOH**

   GFM KOH = (39) + (16) + (1) = 56 g

   $n = \text{mass/GFM} = \frac{1.12}{56} = 0.02\,mol$

   Step 2 – Use the 'concentration' triangle

   $C = \frac{n}{V} = \frac{0.02}{0.25} = 0.08\,mol\,l^{-1}$ $\boxed{\frac{250}{1000} = 0.25;\ \text{volume must be in litres}}$

5. What mass of sodium nitrate ($NaNO_3$) is required to make $100\,cm^3$ of a $2\,mol\,l^{-1}$ solution of sodium nitrate?

   Step 1 – Use the 'concentration' triangle to find the number of moles (n) of $NaNO_3$.

   $n = C \times V = 2 \times 0.1 = 0.2\,mol$ $\boxed{\frac{100}{1000} = 0.1;\ \text{volume must be in litres}}$

   Step 2 – Use the 'mass' triangle

   GFM $NaNO_3$ = (23) + (14) + (3 × 16) = 85 g

   mass = n × GFM = $0.2 \times 85 = 17\,g$

DON'T FORGET

In these examples you will notice that even although the names of the chemicals were given in the question, they were not needed to work out the answers. Many students will needlessly work out the formula and the GFM in this type of question.

ONLINE

Try out the 'Concentration' activity at www.brightredbooks.net/N5Chemistry.

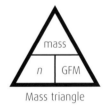

Mass triangle

Concentration triangle

THINGS TO DO AND THINK ABOUT

1 Calculate the concentration, in $mol\,l^{-1}$, of the following solutions:

   (a) $5.85\,g$ of NaCl in $500\,cm^3$ of solution

   (b) $2.0\,g$ of NaOH in 2 litres of solution

   (c) $4.5\,g$ of $C_6H_{12}O_6$ in $100\,cm^3$ of solution

   (d) $2.12\,g$ of Na_2CO_3 in $250\,cm^3$ of solution.

2 Calculate the mass of dissolved solute in the following solutions:

   (a) 2 litres of $LiNO_3$, concentration $0.5\,mol\,l^{-1}$

   (b) $200\,cm^3$ of KI, concentration $2\,mol\,l^{-1}$

   (c) $250\,cm^3$ of HCl, concentration $1\,mol\,l^{-1}$

   (d) $50\,cm^3$ of K_2CO_3, concentration $0.1\,mol\,l^{-1}$.

VIDEO LINK

Check out the video clip 'What mass of salt is needed to make a solution?' at www.brightredbooks.net/N5Chemistry.

FORMULAE AND REACTION QUANTITIES: THE MOLE 3 – EQUATIONS

DON'T FORGET

Chemical equations use arrows (→) not equal signs (=) to separate the reactants and products.

ONLINE

Revise word equations further at www.brightredbooks.net/N5Chemistry.

DON'T FORGET

Diatomic elements are written as X_2. Never write H for hydrogen or Cl for chlorine in a formula equation.

ONLINE TEST

Revise this topic by taking the 'Equations' test at www.brightredbooks.net/N5Chemistry.

WRITING EQUATIONS

When a chemical reaction occurs it can be described by an equation. The equation will show the **reactants** (the starting substances) and the **products** (the substances that are made). The reactants and products are separated by an arrow, with the reactants on the left-hand side and the products on the right-hand side.

reactants → products

The chemicals involved in the reaction can be represented by words or by their chemical formula.

A **word equation** is the simplest type of chemical equation.

Consider the reaction of hydrogen with oxygen to produce hydrogen oxide (water).

The word equation for this reaction is:

hydrogen + oxygen → hydrogen oxide

A **formula equation** shows the correct formulae of all the substances involved in the reaction. The formula equation for this reaction is:

$$H_2 + O_2 \rightarrow H_2O$$

State symbols

Chemicals can exist as solids, liquids or gases. Often reactions take place in aqueous solution (the substance is dissolved in water). State symbols can be added to the right of

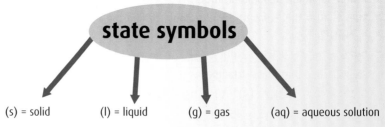

state symbols

(s) = solid   (l) = liquid   (g) = gas   (aq) = aqueous solution

a chemical formula to indicate the state of the substance.

For example, hydrogen gas would be written as $H_2(g)$ and solid magnesium would be written as $Mg(s)$.

BALANCED FORMULA EQUATIONS

In a chemical reaction the atoms present at the start are rearranged into new substances. During this process no atoms are lost or gained. This means that the total mass of the atoms before the reaction must equal the total mass of the atoms when the reaction is complete. A balanced formula equation is one that has the same number and type of atoms on each side of the equation.

contd

EXAMPLE 1

The reaction of hydrogen with oxygen

hydrogen	+	oxygen	→	hydrogen oxide
H_2	+	O_2	→	H_2O

It looks like **an oxygen atom** has been lost – this is **not** possible

This equation is not balanced as there are two oxygen atoms on the left-hand side but only one on the right-hand side. Add an extra water molecule to give two oxygen atoms on each side of the equation:

$$H_2 \quad + \quad O_2 \quad \rightarrow \quad 2H_2O$$

Unfortunately, the equation is still not balanced. There are now two hydrogen atoms on the left-hand side and four on the right. Add an extra hydrogen molecule to the left hand side:

$$2H_2 \quad + \quad O_2 \quad \rightarrow \quad 2H_2O$$

 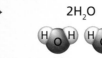

The equation is balanced as the number of atoms on each side is the same.

Always:

- Place the balancing numbers in front of the chemical symbol. For example, $3H_2O$ is acceptable but $H3_2O$ and H_23O are not.
- Carry out a final check to ensure the same number of atoms appear on both sides of the equation.

Never:

- Change a correct chemical formulae. For example, the formula of magnesium chloride is $MgCl_2$. This would not be changed to $MgCl$ or Mg_2Cl for any reason.

EXAMPLE 2

The reaction of magnesium with dilute nitric acid.

magnesium	+	nitric acid	→	magnesium nitrate	+	hydrogen
Mg	+	HNO_3	→	$Mg(NO_3)_2$	+	H_2

ONE nitrate group

TWO nitrate groups

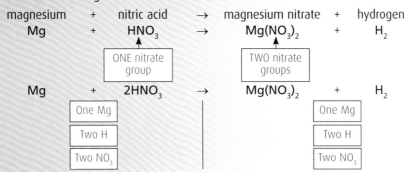

Mg	+	$2HNO_3$	→	$Mg(NO_3)_2$	+	H_2

One Mg | Two H | Two NO_3 | One Mg | Two H | Two NO_3

When groups of atoms like nitrate, carbonate and sulfate appear on both sides of an equation it is often easier to consider the group as a whole. In this case, placing a 2 in front of the nitric acid will balance the nitrate groups. This is much easier than trying to balance the individual oxygen and nitrogen atoms. The 2 also balances the hydrogen atoms.

THINGS TO DO AND THINK ABOUT

Balance the following equations:

(a) $C + F_2 \rightarrow CF_4$

(b) $Cl_2 + KBr \rightarrow KCl + Br_2$

(c) $Al + O_2 \rightarrow Al_2O_3$

(d) $CH_4 + O_2 \rightarrow CO_2 + H_2O$

(e) $FeCl_3 + NaOH \rightarrow Fe(OH)_3 + NaCl$

DON'T FORGET

We cannot add the oxygen atom to the right-hand side of the equation by changing H_2O to H_2O_2. This would change the formula of water. Similarly, we cannot remove an atom of oxygen by changing O_2 to O as O_2 is the formula of oxygen gas.

ONLINE

Practice balancing equations by clicking on 'Balancing Act' at www.brightredbooks.net/N5Chemistry.

DON'T FORGET

Balancing chemical equations can be a matter of trial and error. An initial decision to include a balancing number for a particular atom will often need to be changed as you work through all the atoms in the equation.

VIDEO LINK

Try this out online, click 'Balancing Equations' at www.brightredbooks.net/N5Chemistry.

FORMULAE AND REACTION QUANTITIES: TH MOLE 4 – EQUATIONS AND CALCULATIONS

Space shuttle

VIDEO LINK

Check out the clip 'Calculating Moles from a Balanced Chemical Equation' at www.brightredbooks.net/ N5Chemistry.

ONLINE TEST

Take the test at www.brightredbooks.net/ N5Chemistry.

DON'T FORGET

These calculations will always involve just two of the substances in the equation. It is very important to correctly identify the substances that the question refers to. This is done by selecting the substances for which a mass is given and for which a mass is to be found.

DON'T FORGET

Whichever method you choose, remember to show all your working. This is true for any chemistry calculation. If you are not careful it is easy to make mistakes. An incorrect answer *on its own* will receive no marks. Partial marks will be awarded if correct working is shown.

WHAT DO EQUATIONS TELL US?

Chemical equations tell us what substances are reacting and what substances are produced. Additionally, a balanced chemical equation indicates the mole ratio in which the substances react or in which they are produced. This point is very important in chemistry. It allows chemists to determine the actual quantities of chemicals required in a reaction and how much product will be produced.

The large orange external tank on the space shuttle actually has two smaller tanks inside it. The larger of these contains liquid hydrogen. The smaller tank contains liquid oxygen. The reaction of hydrogen with oxygen produces the power required to lift the shuttle. NASA scientists know that hydrogen reacts with oxygen in a two mole to one mole ratio because they know the balanced equation for the reaction:

$$2H_2 + O_2 \rightarrow 2H_2O$$

Two moles of hydrogen react with one mole of oxygen to produce two moles of water. As a consequence, the tanks are designed to hold at least twice as much hydrogen as oxygen.

MASS CALCULATIONS

You must be able to calculate the mass of a reactant or product in a reaction given the balanced chemical equation.

EXAMPLE 1

Iron(III) oxide reacts with carbon to produce iron and carbon dioxide:

$2Fe_2O_3 + 3C \rightarrow 4Fe + 3CO_2$

Calculate the mass of carbon that will react with 1·6 g of iron(III) oxide.

Step 1 – Select the required chemicals and write down the number of moles of each as they appear in the balanced equation:

$2Fe_2O_3 \leftrightarrow 3C$

2 moles ↔ 3 moles

Step 2 – Convert the given mass into a number of moles.

moles of $Fe_2O_3 = \frac{mass}{GFM} = \frac{1·6}{160} = 0·01$ mol

Step 3 – Use the mole ratio from the equation to find the number of moles of the substance asked for in the question.

2 moles ↔ 3 moles

0·01 ↔ $(\frac{3}{2}) \times 0·01 = 0·015$ mol

Step 4 – Convert the number of moles into mass.

mass of $C = n \times GFM = 0·015 \times 12 = 0·18$ g

This calculation shows that 0·18 g of carbon is needed to react with 1·6 g of iron(III) oxide.

Alternative method for Example 1

Step 1 – Select the required chemicals and write down the number of moles of each as they appear in the balanced equation.

$2Fe_2O_3 \leftrightarrow 3C$

2 moles ↔ 3 moles

Step 2 – Convert the moles into masses.

contd

$2[(2 \times 56) + (3 \times 16)] \leftrightarrow 3 \times 12$

$= 320\,g \qquad\qquad \leftrightarrow\ = 36\,g$

Step 3 – Use the mass given in the question to find the mass of the unknown by simple proportion.

$1{\cdot}6\,g \leftrightarrow \left(\frac{1{\cdot}6}{320}\right) \times 36$

$= 0{\cdot}18\,g$

EXAMPLE 2

Ethanol (C_2H_5OH) burns in oxygen to produce carbon dioxide and water:

$C_2H_5OH + 3O_2 \rightarrow 2CO_2 + 3H_2O$

Calculate the mass of carbon dioxide produced when 11·5 g of ethanol is burned in excess oxygen.

Step 1 – Select the substance for which a mass has been given (ethanol) and for which a mass has to be found (carbon dioxide).

$C_2H_5OH \leftrightarrow 2CO_2$

1 mole $\leftrightarrow$ 2 moles

Step 2 – Calculate the number of moles of ethanol.

$mol = \frac{mass}{GFM} = \frac{11{\cdot}5}{46} = 0{\cdot}25\,mol$

Step 3 – Use the mole ratio from the equation to find the number of moles of carbon dioxide.

1 mole $\leftrightarrow$ 2 moles

$0{\cdot}25 \quad \leftrightarrow 0{\cdot}25 \times 2 = 0{\cdot}5\,mol$

Step 4 – Calculate the mass of carbon dioxide.

$mass = n \times GFM = 0{\cdot}5 \times 44 = 22\,g$

An alternative method for Example 2

$1\,mol \rightarrow 2\,mol$

$46\,g \rightarrow 88\,g$

$11{\cdot}5\,g \rightarrow \frac{(11{\cdot}5 \times 88)}{46}$

$\qquad\qquad = 22\,g$

 DON'T FORGET

Equation calculations will often state that one of the reactants is in excess. Notice that in this example the oxygen was in excess. As the quantities of the other substances in the reaction cannot be determined from the excess reactant, it should be ignored in the calculation.

 DON'T FORGET

Occasionally this type of question may involve kilograms or tonnes of chemical rather than grams. In examples of this kind, the method used in the calculation will not change. However, the final unit will need to be altered. For example, in the ethanol example, using 11·5 kilograms of ethanol would produce 22 kilograms of carbon dioxide and using 11·5 tonnes of ethanol would produce 22 tonnes of carbon dioxide.

 ## THINGS TO DO AND THINK ABOUT

1   Sodium reacts with chlorine to produce sodium chloride.

$2Na + Cl_2 \rightarrow 2NaCl$

   (a) How many moles of sodium react with one mole of chlorine?

   (b) How many moles of sodium chloride are produced from six moles of chlorine?

   (c) How many moles of sodium chloride are produced from five moles of sodium?

2   Magnesium reacts with dilute hydrochloric acid to produce hydrogen gas.

$Mg + 2HCl \rightarrow MgCl_2 + H_2$

Calculate the mass of magnesium needed to produce 100 g of hydrogen.

3   Methane (CH_4) is a major component of natural gas. Calculate the mass of carbon dioxide produced when 32 kilograms of methane are burned.

$CH_4 + 2O_2 \rightarrow CO_2 + 2H_2O$

 ONLINE

Try out more calculations at www.brightredbooks.net/N5Chemistry.

FORMULA AND REACTION QUANTITIES: THE MOLE 5 – PERCENTAGE COMPOSITION

WHAT IS PERCENTAGE COMPOSITION?

The percentage composition of a compound is a way of describing the proportion of different elements contained in the compound and is simply the percentage by mass of each element in the compound.

Percentage composition has important uses in industry.

In agriculture it provides important information regarding the quantity of essential elements in a fertiliser.

nitrogen   phosphorus   potassium
(N) 20%   (P) 5%   (K) 10%

This packaging shows the numbers 20-5-10, which are the percentage compositions for each of the essential elements N, P and K

When metals are extracted from their ores the percentage composition of the ore allows chemists to determine the expected mass of metal contained in a given mass of the metal ore.

Iron oxide, Fe_2O_3 – 70% iron

CALCULATING PERCENTAGE COMPOSITION

To calculate the percentage by mass of an element in a compound, the chemical formula of the compound must be known. The formula allows the **total mass** of the element in the compound together with the **gram formula mass** (GFM) of the compound to be calculated.

Both these masses are substituted into the relationship shown below.

Percentage by mass of element $= \dfrac{\text{total mass of element}}{\text{GFM of compond}} \times 100$

DON'T FORGET

Now would be a good time to revise gram formula mass. This can be found on page 34 of the Study Guide.

EXAMPLE 1

Calculate the percentage by mass of potassium in potassium carbonate, K_2CO_3.

Step 1 total mass of potassium $= 2 \times 39 = 78$

Step 2 gram formula mass of potassium carbonate
$= (2 \times 39) + (12) + (3 \times 16) = 138$

Step 3 % K $= \dfrac{78}{138} \times 100 = 56\cdot5\%$

In this example the mass of potassium used in the relationship was 78. This is the **total mass of potassium in the compound** (K_2CO_3 gives $2 \times K = 2 \times 39 = 78$).

Many students will incorrectly use 39 as the total mass of potassium in this calculation. Always remember to refer to the formula of the compound to make sure that the total mass of the element is used.

ONLINE

Follow the links on the Digital Zone for more on percentage composition.

EXAMPLE 2

Calculate the percentage by mass of nitrogen in ammonium carbonate, $(NH_4)_2CO_3$.

Step 1 total mass of nitrogen $= 2 \times 14 = 28$

Step 2 gram formula mass of ammonium carbonate
$= (2 \times 14) + (8 \times 1) + (12) + (3 \times 16) = 96$

Step 3 % N $= \dfrac{28}{96} \times 100 = 29\cdot2\%$

In this example the compound is composed of four different elements: nitrogen, hydrogen, carbon and oxygen. Many students will incorrectly calculate the percentage mass of each element as 25% believing that each elements' mass makes up one quarter of the mass of the compound. This is not the case as each element has a different relative formula mass. Do not make this mistake in the National 5 exam.

VIDEO LINK

Learn more about percentage composition by following the clips at www.brightredbooks.net

THINGS TO DO AND THINK ABOUT

1 Calculate the percentage by mass of sodium in sodium sulfate, Na_2SO_4.

2 Calculate the percentage by mass of carbon in ethanol, C_2H_6O.

3 Calculate the percentage by mass of magnesium in magnesium nitrate, $Mg(NO_3)_2$.

4 Calculate the percentage by mass of calcium in calcium carbonate, $CaCO_3$.

5 Calculate the percentage by mass of titanium in titanium dioxide, TiO_2.

ONLINE TEST

Test yourself on this topic on the Digital Zone.

ACIDS AND ALKALIS

THE DISSOCIATION OF WATER

A solution of sodium chloride, Na^+Cl^-(aq), like all ionic solutions, is a good conductor of electricity. In a previous topic on bonding we found out that this was due to the ions in the solution being free to move.

Pure water is a very poor conductor of electricity. But the fact that it does conduct, even to a very slight extent, shows that ions must be present.

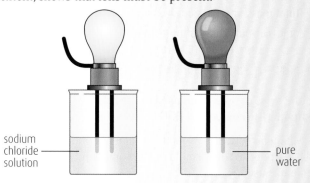

sodium chloride solution

pure water

Salt water has a conductivity that is 1 000 000 times greater than pure water. Ions are present in pure water because a very small proportion, about one in every 500 million water molecules, H_2O(l), break up into hydrogen ions, H^+(aq) and hydroxide ions, OH^-(aq). The scientific term **dissociate** is usually used instead of break up.

The dissociation of water molecules is represented by the equation:

$$H_2O(l) \rightleftharpoons H^+(aq) + OH^-(aq)$$

It is clear from the equation that when a water molecule dissociates it will produce one hydrogen ion and one hydroxide ion. This means that in pure water the **concentration of hydrogen ions is equal to the concentration of hydroxide ions**.

DON'T FORGET

The (aq) symbol indicates that the substance is dissolved in water.

DON'T FORGET

A double-headed arrow in an equation means that the reaction can occur in both directions. The reaction is **reversible**.

THE pH SCALE

The **pH scale** is a measure of how **acidic** a solution is and is a continuous range from below 0 to above 14. The pH of a solution can be found by adding pH paper or a pH indicator such as universal indicator to the solution and matching the colour of the paper or indicator to a pH colour chart like the one shown here.

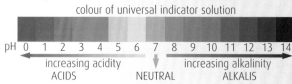

colour of universal indicator solution

pH 0 1 2 3 4 5 6 7 8 9 10 11 12 13 14

increasing acidity
ACIDS

NEUTRAL

increasing alkalinity
ALKALIS

ONLINE

Get some extra revision on acids and alkalis at www.brightredbooks.net/N5Chemistry.

ONLINE

Explore the pH scale further online at www.brightredbooks.net/N5Chemistry.

- Any substance with a pH value less than 7 is an **acid**. The lower the number the more acidic the substance is.

- Any substance with a pH value greater than 7 is an **alkali**. The higher the number the more alkaline the substance is.

- Pure water and neutral solutions have a pH equal to 7.

- The sulfuric acid in a car battery has a pH of less than 1 and is much more acidic than vinegar, which has a pH of 3.

- Caustic soda, a solution of sodium hydroxide, has a pH of 14 and is much more alkaline than a baking soda solution, which typically has a pH of 8 or 9.

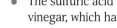 contd

pH, hydrogen ions (H⁺(aq)) and hydroxide ions (OH⁻(aq))

All aqueous solutions will contain both hydrogen ions and hydroxide ions. It is the relative concentrations of these ions that decide if a solution is acidic, alkaline or neutral. The **pH number** of any aqueous solution is actually a measure of the **concentration of hydrogen ions** the solution contains. The greater the concentration of hydrogen ions a solution contains, the lower its pH will be, and vice versa.

In an acid solution the concentration of hydrogen ions is greater than the concentration of hydroxide ions.

$H^+(aq) > OH^-(aq)$ **Acid**

In an alkaline solution the concentration of hydroxide ions is greater than the concentration of hydrogen ions.

$OH^-(aq) > H^+(aq)$ **Alkali**

In pure water and all neutral solutions the concentration of hydrogen ions is equal to the concentration of hydroxide ions.

$OH^-(aq) = H^+(aq)$ **Neutral**

ONLINE

For more resources on the pH scale, visit www.brightredbooks.net/N5Chemistry.

DON'T FORGET

Acidic solutions have a higher concentration of hydrogen ions [H⁺(aq)] than pure water while alkaline solutions have a higher concentration of hydroxide ions [OH⁻(aq)] than pure water.

DILUTING ACIDS AND ALKALIS

The addition of water to any aqueous solution will **decrease the concentration** of the solution – the solution will become more dilute. In the lab we can dilute an acidic or alkaline solution with water and measure the pH of the solution with a pH meter.

Typical results from dilution experiments are shown in the following tables.

Diluting 1 mol l⁻¹ hydrochloric acid, HCl(aq)

Concentration (mol l⁻¹)	1	0·1	0·01	0·001	0·0001	0·00001	0·000001	0·0000001
pH	0	1	2	3	4	5	6	7

As the acid solution is diluted the concentration of **hydrogen ions decreases**. This causes the **pH to increase towards 7.**

Diluting 1 mol l⁻¹ sodium hydroxide, NaOH(aq)

Concentration (mol l⁻¹)	1	0·1	0·01	0·001	0·0001	0·00001	0·000001	0·0000001
pH	14	13	12	11	10	9	8	7

As the alkaline solution is diluted the concentration of **hydroxide ions decreases**. This causes the **pH to decrease towards 7.**

VIDEO LINK

Check out the video on 'Acids and Alkalis' at www.brightredbooks.net/N5Chemistry.

DON'T FORGET

Students often get confused when asked questions about diluting acids or alkalis. An alkaline pH will fall towards 7. An acid pH will rise towards 7. The concentration of the ions will always decrease when the solution is diluted.

THINGS TO DO AND THINK ABOUT

1   Consider the data in the table.

(a) Identify the solutions that have a higher concentration of hydrogen ions than hydroxide ions.

(b) Identify the solution that has an equal concentration of hydrogen ions and hydroxide ions.

Solution	pH
Vinegar	3
Baking soda	9
Sodium chloride	7
Lemonade	2

2   This question appeared in a chemistry exam: 'Explain what happens to the pH of an alkaline solution when it is diluted with water'. The answers given by four students were:

A   The pH goes to 7 because of the water.

B   The amount of hydrogen goes down and so the pH goes down.

C   The concentration of the alkali decreases and so the pH becomes 7.

D   As water is added the pH goes down because it is more dilute.

(a) Discuss these answers with your classmates and decide which, if any, are correct, partially correct or incorrect.

(b) Can your group come up with a better answer than those given by the four students?

ONLINE TEST

Check your knowledge of the pH scale online at www.brightredbooks.net/N5Chemistry.

ACIDS AND BASES: MAKING ACIDS AND ALKALIS

THE FORMULAE OF ACIDS AND ALKALIS

Acids and alkalis are much better conductors of electricity than pure water due to the fact that they are ionic compounds. Solutions of acids and alkalis contain ions that are free to move and so carry an electric current.

The following tables show the ions present in some common laboratory acids and alkalis and the various ways to write the formulae of aqueous solutions of these acids and alkalis.

Acid	Ions present	Formula without charges	Formula showing charges	Expanded formula showing separate ions
hydrochloric	hydrogen ions and chloride ions	$HCl(aq)$	$H^+Cl^-(aq)$	$H^+(aq) + Cl^-(aq)$
nitric	hydrogen ions and nitrate ions	$HNO_3(aq)$	$H^+NO_3^-(aq)$	$H^+(aq) + NO_3^-(aq)$
sulfuric	hydrogen ions and sulfate ions	$H_2SO_4(aq)$	$(H^+)_2SO_4^{2-}(aq)$	$2H^+(aq) + SO_4^{2-}(aq)$

Alkali	Ions present	Formula without charges	Formula showing charges	Expanded formula showing separate ions
sodium hydroxide	sodium ions and hydroxide ions	$NaOH(aq)$	$Na^+OH^-(aq)$	$Na^+(aq) + OH^-(aq)$
potassium hydroxide	potassium ions and hydroxide ions	$KOH(aq)$	$K^+OH^-(aq)$	$K^+(aq) + OH^-(aq)$
calcium hydroxide	calcium ions and hydroxide ions	$Ca(OH)_2(aq)$	$Ca^{2+}(OH^-)_2(aq)$	$Ca^{2+}(aq) + 2OH^-(aq)$

Non-metal oxides

Acids are formed when soluble non-metal oxides dissolve in water.

EXAMPLE

Sulfur dioxide gas, $SO_2(g)$, dissolves in water to form sulfurous acid:

sulfur dioxide	+	water	→	sulfurous acid
$SO_2(g)$	+	$H_2O(l)$	→	$H_2SO_3(aq)$

Sulfurous acid is one of the components of acid rain.

VIDEO LINK

Watch the clip 'Burning Sulphur' at www.brightredbooks.net/N5Chemistry.

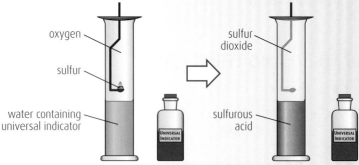

oxygen

sulfur

water containing universal indicator

sulfur dioxide

sulfurous acid

In this experiment yellow sulfur is burned in oxygen gas to produce sulfur dioxide gas. The sulfur dioxide gas formed dissolves in the water, turning the universal indicator red and confirming that an acid has been produced.

EXAMPLE

Carbon dioxide, $CO_2(g)$, dissolves in water to form carbonic acid:

carbon dioxide	+	water	→	carbonic acid
$CO_2(g)$	+	$H_2O(l)$	→	$H_2CO_3(aq)$

Carbonic acid is found in all carbonated (fizzy) drinks.

contd

The hydrogen ion, $H^+(aq)$, concentration of water will increase when a soluble non-metal oxide dissolves in the water. The expanded ionic formulae of sulfurous acid and carbonic acid show the presence of hydrogen ions:

sulfurous acid $\quad 2H^+(aq) + SO_3^{2-}(aq)$

carbonic acid $\quad 2H^+(aq) + CO_3^{2-}(aq)$

Metal oxides

Alkalis are formed when soluble metal oxides or hydroxides dissolve in water. Some metal oxides react with water to form metal hydroxides.

EXAMPLE

Sodium oxide, $Na_2O(s)$, reacts with water to form sodium hydroxide solution.

sodium oxide + water → sodium hydroxide

$Na_2O(s) \quad + H_2O(l) → 2NaOH(aq)$

Sodium hydroxide is a strong alkali often referred to as caustic soda. It is used in many household cleaning products.

EXAMPLE

Calcium oxide, $CaO(s)$, reacts with water to form calcium hydroxide solution.

calcium oxide + water → calcium hydroxide

$CaO(s) \quad + H_2O(l) → Ca(OH)_2(aq)$

Calcium hydroxide is an alkali which is used to raise the pH of acidic soils.

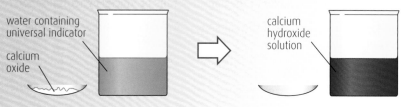

water containing universal indicator

calcium oxide

calcium hydroxide solution

In this experiment solid calcium oxide powder is added to some water containing universal indicator. The calcium oxide dissolves in the water, turning the universal indicator blue and confirming the presence of an alkali.

The hydroxide ion $OH^-(aq)$, concentration of water will increase when a soluble metal oxide is dissolved in the water. The expanded ionic formulae of sodium hydroxide and calcium hydroxide show the presence of hydroxide ions:

sodium hydroxide $\quad Na^+(aq) + OH^-(aq)$
calcium hydroxide $\quad Ca^{2+}(aq) + 2OH^-(aq)$

VIDEO LINK

Have a look at the 'Basic Oxide Hydration' clip at www.brightredbooks.net/N5Chemistry.

DON'T FORGET

To produce an acidic or alkaline solution an oxide must be soluble in water. Adding an insoluble oxide to water will leave the pH of the water unchanged. The data booklet gives information about the solubility of a variety of compounds.

ONLINE TEST

Take the 'Making Acids and Alkalis' test online at www.brightredbooks.net/N5Chemistry.

THINGS TO DO AND THINK ABOUT

1 Jennifer added some copper(II) oxide, calcium oxide and aluminium oxide to three separate test tubes. She added some water to the test tubes and then tested the pH of the water in each. Jennifer's results are shown in the table.

    (a) Explain why the pH reading was 7 for two of the oxides.

    (b) Write a balanced equation to show what happened to the calcium oxide (CaO) when it was added to the water (H_2O).

Chemical	pH
copper(II) oxide	7
calcium oxide	9
aluminium oxide	7

2 When the element bromine is added to some water the following reaction takes place:

$Br_2 + H_2O → 2H^+ + Br^- + BrO^-$

Will the pH of the water be lower or higher than 7 when bromine is added? Explain your answer.

ACIDS AND BASES: NEUTRALISATION REACTIONS

DON'T FORGET

The terms base and alkali are often confused by students. All alkalis are bases but not all bases are alkalis. For example calcium carbonate is a base but as it does not dissolve in water it cannot be classified as an alkali. Sodium hydroxide is a base and because it is soluble it forms an alkali when dissolved in water.

WHAT IS A BASE?

A **base** is a substance that **neutralises** an acid. There are three main types of base: metal oxides, metal hydroxides and metal carbonates. If a base is soluble in water it forms an **alkali**. Burning magnesium produces the base magnesium oxide, MgO. The base calcium hydroxide, $Ca(OH)_2$, is also known as lime. The base strontium carbonate, $SrCO_3$, produces the bright red colour in fireworks.

WHAT IS A NEUTRALISATION REACTION?

When a base is added to an **acid** a neutralisation reaction takes place. In a neutralisation reaction the pH of the acid will rise towards 7 as the concentration of hydrogen ions decreases. Water and new ionic substances called **salts** are always formed during this process.

We can deduce the salt formed in a neutralisation reaction from the names of the acid and base involved in the reaction. Salts formed from hydrochloric acid (HCl) are called chlorides, sulfuric acid (H_2SO_4) are called sulfates and nitric acid (HNO_3) are called nitrates.

The table shows some examples of the names of salts produced in neutralisation reactions.

You can see from the names of the salts that the first part of the name comes from the first part of the name of the base.

Base	Acid	Salt
sodium hydroxide	hydrochloric	sodium chloride
magnesium oxide	sulphuric	magnesium sulfate
lithium carbonate	nitric	lithium nitrate

Acids and metal oxides

When a metal oxide reacts with an acid a salt and water are formed:

metal oxide + acid → salt + water

> **EXAMPLE**
>
> Copper(II) oxide reacts with dilute sulfuric acid forming copper(II) sulfate solution and water.
>
> Word equation: copper(II) oxide + sulfuric acid → copper(II) sulfate + water
>
> Formula equation: $CuO(s) + H_2SO_4(aq) → CuSO_4(aq) + H_2O(l)$
>
> Formula equation showing ions:
> $Cu^{2+}O^{2-}(s) + 2H^+(aq) + SO_4^{2-}(aq) → Cu^{2+}(aq) + SO_4^{2-}(aq) + H_2O(l)$

VIDEO LINK

Check out the 'Neutralisation of sulfuric acid with copper oxide' clip at www.brightredbooks.net/N5Chemistry.

Notice that the sulfate ion and the copper(II) ion are exactly the same on either side of the equation. This means that even although these ions are part of the reaction mixture they have not been chemically changed. Ions like this are called **spectator ions**. Writing the equation without the spectator ions shows only the reacting species:

$2H^+(aq) + O^{2-}(s) → H_2O(l)$

This equation shows that any metal oxide will react with any acid to produce water.

In the laboratory the preparation of crystals of the salt copper(II) sulfate using the reaction described above could be carried out as shown.

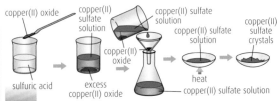

copper(II) oxide
copper(II) sulfate solution
copper(II) sulfate solution
copper(II) sulfate solution
copper(II) sulfate crystals
copper(II) oxide
sulfuric acid
excess copper(II) oxide
heat
copper(II) sulfate solution

Copper(II) oxide is added to the dilute sulfuric acid until it stops dissolving (gentle heat is applied at this stage to speed up the reaction). At this point all the acid in the beaker has been neutralised. As copper(II) oxide is insoluble, the excess copper(II) oxide is removed by filtration. The copper(II) sulfate solution is heated and the water evaporates off leaving crystals of copper(II) sulfate.

contd

Acids and alkalis

Alkalis react with acids to form a salt and water:

acid + alkali → salt + water

> **EXAMPLE**
>
> Sodium hydroxide solution reacts with dilute hydrochloric acid forming sodium chloride solution and water.
>
> Word equation: sodium hydroxide + hydrochloric acid → sodium chloride + water
>
> Formula equation: $NaOH(aq) + HCl(aq) \rightarrow NaCl(aq) + H_2O(l)$
>
> Formula equation showing ions:
>
> $Na^+(aq) + OH^-(aq) + H^+(aq) + Cl^-(aq) \rightarrow Na^+(aq) + Cl^-(aq) + H_2O(l)$
>
> Notice that the sodium ion and the chloride ion are spectator ions in this reaction. Writing the equation without the spectator ions shows the reacting species.
>
> $H^+(aq) + OH^-(aq) \rightarrow H_2O(l)$

VIDEO LINK

For more, go online and watch the 'Neutralisation Reactions' video at www.brightredbooks.net/N5Chemistry.

This equation shows that when any acid reacts with any alkali the only chemical change is the reaction between the hydrogen ions from the acid and the hydroxide ions from the alkali to form water.

In the laboratory, the preparation of crystals of the salt sodium chloride using the reaction described above could be carried out as shown.

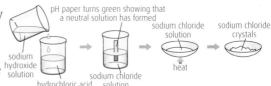

pH paper turns green showing that a neutral solution has formed

sodium hydroxide solution

hydrochloric acid

sodium chloride solution

heat

sodium chloride solution

sodium chloride crystals

Sodium hydroxide solution is added to the dilute hydrochloric acid until the solution is neutral. This can be checked using pH paper, as there would be no visible signs in the beaker to indicate this. There is no need to filter the solution as there are no insoluble chemicals present. The sodium chloride solution is heated to evaporate the water, leaving sodium chloride crystals.

Acids and metal carbonates

Metal carbonates react with acids to form a salt and water and carbon dioxide:

metal carbonate + acid → salt + water + carbon dioxide

> **EXAMPLE**
>
> Calcium carbonate reacts with dilute nitric acid forming calcium nitrate solution, water and carbon dioxide.
>
> Word equation:
>
>   calcium carbonate + nitric acid → calcium nitrate + water + carbon dioxide
>
> Formula equation:
>
> $CaCO_3(s) + 2HNO_3(aq) \rightarrow Ca(NO_3)_2(aq) + H_2O(l) + CO_2(g)$
>
> Formula equation showing ions:
>
> $Ca^{2+}CO_3^{2-}(s) + 2H^+(aq) + 2NO_3^-(aq) \rightarrow Ca^{2+}(aq) + 2NO_3^-(aq) + H_2O(l) + CO_2(g)$
>
> Notice that the calcium ion and the nitrate ion are spectator ions in this reaction. Writing the equation without the spectator ions shows the reacting species:
>
> $2H^+(aq) + CO_3^{2-}(s) \rightarrow H_2O(l) + CO_2(g)$

This equation shows that any acid reacts with any carbonate to form water and carbon dioxide.

VIDEO LINK

See how this works in practice by watching 'Carbonate Reactions in Medicine' at www.brightredbooks.net/N5Chemistry.

THINGS TO DO AND THINK ABOUT

Classify the following chemicals as bases, alkalis or salts.

(a) Sodium sulfate

(b) Potassium hydroxide

(c) Iron(III) chloride

(d) Zinc(II) oxide

(e) Sodium oxide

ACIDS AND BASES: TITRATIONS

HOW TO CARRY OUT A TITRATION

Titration is an analytical technique used to determine the accurate concentration of a solution by reacting it with a solution of known concentration. Suppose you wanted to find the concentration of a sodium hydroxide solution given a $0.25\,mol\,l^{-1}$ solution of sulfuric acid. A possible experimental procedure is detailed below.

Use a pipette to measure out $25\,cm^3$ of sodium hydroxide solution into a conical flask.

Add two drops of pH indicator to the flask.

Rinse and fill a burette with $0.25\,mol\,l^{-1}$ sulfuric acid.

Record the initial reading on the burette and run the acid into the flask until the indicator changes colour.

Record the final burette reading.

Using a pipette

Normally, an initial titration is carried out quickly to determine the rough volume of acid required to reach the end-point. The titration is then repeated much more carefully, with the acid being added drop by drop as the end-point is approached, until at least two volumes are within $0.2\,cm^3$ of each other. The volumes are then said to be **concordant**.

DON'T FORGET

pH indicators change colour at the end-point of a titration. This is the point where the reaction of the acid and alkali is complete.

Carrying out a titration

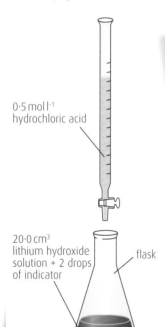

$0.5\,mol\,l^{-1}$
hydrochloric acid

$20.0\,cm^3$
lithium hydroxide
solution + 2 drops
of indicator

flask

CALCULATIONS BASED ON TITRATIONS

EXAMPLE

Using dilute hydrochloric acid, Brenda carried out titrations to determine the concentration of a solution of lithium hydroxide.

	Rough titre	1st titre	2nd titre
Initial burette reading (cm³)	0·3	0·2	0·5
Final burette reading (cm³)	26·6	25·3	25·4
Volume used (cm³)	26·3	25·1	24·9

(a) Why was indicator added to the flask?

Indicators are added to determine the end-point of the titration.

(b) What average volume should be used in calculating the concentration of the lithium hydroxide solution?

You must take only the concordant titre values to calculate the average volume:

average volume = $\frac{25\cdot1 + 24\cdot9}{2}$ = $25\cdot0\,cm^3$

The rough titre of $26\cdot3\,cm^3$ is not included in the average volume.

(c) The equation for the reaction between hydrochloric acid and lithium hydroxide is:

$HCl(aq) + LiOH(aq) \rightarrow LiCl(aq) + H_2O(l)$

Calculate the concentration of the lithium hydroxide solution.

Step 1

In part (b) you calculated the average volume ($V = 25\cdot0\,cm^3 = 0\cdot025$ litres) of the hydrochloric acid. The concentration ($C = 0\cdot50\,mol\,l^{-1}$) of the hydrochloric acid is shown next to the burette. This allows you to determine the number of moles (n) of hydrochloric acid needed to neutralise the lithium hydroxide in the flask.

HCl: $n = C \times V = 0\cdot50 \times 0\cdot025 = 0\cdot0125\,mol$

DON'T FORGET

It is easy to make mistakes when asked to calculate the average volume of acid (or alkali) used in a titration. It is only those volumes that are concordant (within $0.2\,cm^3$ of each other) that are used to calculate the average volume.

contd

Step 2

You can use the balanced equation to work out the number of moles of lithium hydroxide that will react with 0·0125 mol of hydrochloric acid.

$$HCl(aq) \leftrightarrow LiOH(aq)$$
$$1\,mol \leftrightarrow 1\,mol$$

So: 0·0125 mol ↔ 0·0125 mol

Step 3

Knowing the number of moles of lithium hydroxide (n = 0·0125 mol) and the volume of the lithium hydroxide (V = 20 cm³ = 0·020 litres), the concentration of the lithium hydroxide can be found.

LiOH (aq): $C = \frac{n}{V} = \frac{0·0125}{0·020} = 0·625\,mol\,l^{-1}$

EXAMPLE

In a titration, it was found that 20 cm³ of potassium hydroxide solution was neutralised by 0·04 mol l⁻¹ of phosphoric acid. The volumes of phosphoric acid used in the titration are shown in the table.

Titration	Volume of 0·04 mol l⁻¹ phosphoric acid (cm³)
1	31·0
2	29·4
3	29·6

(a) **Using the results in the table, calculate the average volume of phosphoric acid required to neutralise the potassium hydroxide solution.**

Taking only the results for titrations 2 and 3:

average volume of phosphoric acid used = $\frac{29·4 + 29·6}{2}$ = 29·5 cm³

(b) **Calculate the number of moles of phosphoric acid used in the titration.**

Knowing the volume (V = 29·5 cm³ = 0·0295 litres) and the concentration (C = 0·04 mol l⁻¹) allows you to work out the number of moles (n) of phosphoric acid.

Phosphoric acid: $n = C \times V = 0·04 \times 0·0295 = 0·00118\,mol$

(c) **Given that one mole of phosphoric acid reacts with three moles of potassium hydroxide, calculate the concentration of the potassium hydroxide solution.**

Notice that this question does not show the equation for the neutralisation reaction but it does give the mole ratio of the two reactants. You can use this mole ratio to work out the number of moles of potassium hydroxide that will react with 0·00118 mol of phosphoric acid:

phosphoric acid ↔ potassium hydroxide

1 mol ↔ 3 mol

0·00118 mol ↔ 0·00118 × 3 = 0·00540 mol

Knowing the number of moles of alkali (n = 0·00540 mol) and the volume of the alkali (V = 20 cm³ = 0·020 litres), the concentration of the alkali can be found.

KOH(aq): $C = \frac{n}{V} = \frac{0·00540}{0·020} = 0·270\,mol\,l^{-1}$

ONLINE

Check out the 'Acid–Base Titration Animation' link at www.brightredbooks.net/N5Chemistry.

VIDEO LINK

Watch the clip 'Setting up and performing a titration' at www.brightredbooks.net/N5Chemistry.

ONLINE TEST

Take the 'Titrations' test online at www.brightredbooks.net/N5Chemistry.

THINGS TO DO AND THINK ABOUT

To find the concentration of a solution of potassium hydroxide Jill titrated 10 cm³ of the alkali with a 0·75 mol l⁻¹ solution of sulfuric acid. The titration was carried out four times and the volumes of sulfuric acid required to neutralise the potassium hydroxide were 40·0 cm³, 38·6 cm³, 37·4 cm³ and 37·3 cm³.

(a) What average volume of sulfuric acid should Jill use to calculate the concentration of the potassium hydroxide solution?

(b) The equation for the neutralisation reaction is: 2KOH(aq) + H₂SO₄(aq) → Na₂SO₄(aq) + 2H₂O(l)

Calculate the concentration of the potassium hydroxide solution.

NATURE'S CHEMISTRY

HOMOLOGOUS SERIES: ALKANES

HYDROCARBONS

Hydrocarbons are compounds that contain carbon and hydrogen only. They can be obtained by **fractional distillation** of crude oil and are the main compounds present in **fossil fuels** such as **natural gas** and **oil**.

Hydrocarbons can be divided into different subsets, sometimes called families. This course considers three of these families – alkanes, alkenes and cycloalkanes.

ALKANE FAMILY

Butane

Most of the compounds found in crude oil-based fuels such as petrol belong to the **alkane** family. Alkanes are commonly used as fuels and are insoluble in water. One member of this family is butane. You may have heard of butane before as it is a common fuel used for portable heaters, barbecuing and in some camping stoves.

In three dimensions butane can be represented as shown.

The grey spheres represent carbon atoms and the white spheres represent hydrogen atoms. This means that butane has four carbon atoms and ten hydrogen atoms, and so its **molecular formula** can be written C_4H_{10}.

All of the carbon atoms are joined together by single covalent bonds and so butane is a **saturated hydrocarbon**.

DIFFERENT TYPES OF FORMULA

As well as representing the three-dimensional structure of alkanes as shown above, there are three other types of formula that can be used to represent hydrocarbons.

Molecular formulae

This is the simplest representation of the compound showing the symbol for each type of atom present and the number of each type.

The molecular formula for butane is: C_4H_{10}.

Full structural formula

The **full structural formula** for butane is:

This shows all the atoms and all the bonds in the molecule.

Shortened structural formula

The **shortened structural formula** for butane can be written in two ways:

$$CH_3-CH_2-CH_2-CH_3 \quad \text{or} \quad CH_3CH_2CH_2CH_3$$

This is an abbreviated version of the full structural formula.

DON'T FORGET

It is useful to know these prefixes and their meanings but you can also find the names of the first eight alkanes listed in numerical order of carbon atoms in the data book.

ONLINE

Go online and take the test on the naming of alkanes at www.brightredbooks.net/N5Chemistry.

NAMING STRAIGHT-CHAIN ALKANES

All hydrocarbons can be named using a systematic naming system. In this system a **prefix** is used that lets other chemists know how many carbons are in the compound. The prefix for butane is 'but-' and this means four.

Prefix	Number of carbons
meth-	1
eth-	2
prop-	3
but-	4
pent-	5
hex-	6
hept-	7
oct-	8

NAMING BRANCHED-CHAIN ALKANES

The alkanes that we have considered up to now have been straight-chain alkanes with all of the carbon atoms joined in a continuous chain. Look at the compound shown on the right:

This alkane has a continuous chain of five carbon atoms (shown in red on the diagram) and so it is based on pent-. Both of the branches contain only one carbon – CH_3 – and this is called a **methyl** group. A branch with two carbon atoms in it would be an **ethyl** group – CH_2CH_3.

To name branched-chain alkanes we need to follow a set of internationally agreed rules:

- Pick out the longest continuous chain of carbon atoms. This forms the basis for the name. In the example above, the longest chain contains the five carbon atoms that are shown in red. In this example the name will be based on pent-.

- Because all of the C–C bands are single, add 'ane' to the name.

- Number this carbon chain – this will allow us to identify where in the chain the branch appears. The chain is always numbered starting from whichever end is closer to the branch. In this case, numbering would start at the left end of the chain.

- Identify the branches and arrange them in alphabetical order. Ethyl would come before methyl and so on. If there are two or more of the same type of branch this is shown in the name by using di, tri, tetra and so on. We also need to indicate, by using a number, which carbon of the main chain the branches are joined to. In the example above there are two methyl branches, one on carbon 2 and one on carbon 3 and so these would be named as 2,3-dimethyl.

- Add 2,3-dimethyl to the base name, so this compound is 2,3-dimethylpentane.

Similar to straight-chain alkanes, branched-chain alkanes are used as fuels.

full structural formula

shortened structural formula

DON'T FORGET

Numbers in a name are separated by commas (,) and numbers and words are separated by hyphens (-).

WRITING STRUCTURAL FORMULAE

You will also need to be able to write a structural formula for a branched-chain alkane from its name. For example, 3-ethylhexane:

- The base of the name is hex- so six carbons will be written in a continuous chain.

- An ethyl ($-CH_2CH_3$) group will be attached to the third carbon atom of the chain.

- The '-ane' ending tells us that all of the carbon-to-carbon bonds are single bonds and as all of the carbon atoms need to have four bonds then the hydrogen atoms are written in.

The shortened structural formula for this compound can be written as:

$CH_3CH_2CH(CH_2CH_3)CH_2CH_2CH_3$ or

ONLINE

Try naming branched chain alkanes and swapping between three different formula types online at www.brightredbooks.net/N5Chemistry.

ONLINE

Do the 'Alkane Quiz' at www.brightredbooks.net/N5Chemistry.

THINGS TO DO AND THINK ABOUT

For the compounds **i–vi** write:

(a) the full structural formula

(b) the shortened structural formula

(c) the molecular formula.

i heptane	**iii** 2-methylpropane	**v** 2,2,4-trimethylpentane
ii propane	**iv** 3,6 -dimethylheptane	**vi** 3-ethyl-4-methylheptane

HOMOLOGOUS SERIES: ALKENES

DON'T FORGET

Many **fractions** are obtained during the distillation of crude oil. There is a very high demand for some fractions whereas other fractions are less useful. To meet demand, fractions containing large, less useful hydrocarbon molecules can be cracked into smaller, more useful molecules. Alkene molecules are also made in this process.

ALKENES: AN INTRODUCTION

Alkenes are another family of hydrocarbons. They are obtained by **cracking crude oil** fractions and can be used to make ethanol and other alcohols and polymers (plastics). Like alkanes, alkenes are insoluble in water.

Alkenes differ from alkanes in that they contain a carbon-to-carbon double covalent bond (C=C) and so are **unsaturated** molecules. The C=C double bond is the **functional group** – the group of atoms in a molecule that are responsible for the chemical reactions that the molecule will undergo.

DON'T FORGET

Alkenes contain a carbon-to-carbon double bond. The first member of the alkene family is ethene as there must be at least two carbon atoms in the molecule.

ALKENE FAMILY

Take a look at the table below containing formulae of the first four members of the alkene family.

Alkene	Full structural formula	Shortened structural formula	Molecular formula
ethene	*(structure shown)*	CH_2CH_2 or $CH_2{=}CH_2$	C_2H_4
propene	*(structure shown)*	CH_3CHCH_2 or $CH_3{-}CH{=}CH_2$	C_3H_6
butene	*(structure shown)*	$CH_3CH_2CHCH_2$ or $CH_3{-}CH_2{-}CH{=}CH_2$	C_4H_8
pentene	*(structure shown)*	$CH_3CH_2CH_2CHCH_2$ or $CH_3{-}CH_2{-}CH_2{-}CH{=}CH_2$	C_5H_{10}

STRAIGHT-CHAIN ALKENES

In the table above the carbon-to-carbon double bond is positioned at the end of the chain. For butene and pentene the C=C could be positioned between different carbon atoms.

If we look at pentene we can see that the C=C can be in two different positions:

A   B

A and **B** *(structural formulae shown)*

DON'T FORGET

Molecules can be in any position in three dimensions. This means that these are the only two forms of straight-chain pentene molecules. If the double bond in the molecule **B** is moved another carbon along in the chain to make –C–C=C–C–C– all that is created is the molecule **A** again but flipped 180°.

Naming straight-chain alkenes

The two structures of pentene shown above have slightly different properties. The boiling point of **A** is 30°C and the boiling point of **B** is 36°C, so they cannot both be called pentene.

To differentiate the two molecules by name, an internationally agreed set of rules is used:

- The number of carbon atoms in the chain is counted and this will determine the prefix (five carbons = pent-).

contd

- The carbon atoms in the chain are numbered, beginning with the end carbon nearest to the double bond. The molecules **A** and **B** would be numbered as shown below.

  A $C^5–C^4–C^3–C^2=C^1$ and **B** $C^5–C^4–C^3=C^2–C^1$

- The position of the C=C bond is given by the first carbon in the double bond. In molecule **A** this is carbon 1 and in molecule **B** this is carbon 2.

- The number is inserted into the name using hyphens (-) to separate the words and numbers.

- The ending '-ene' is added. This shows that the compound is a member of the alkene homologous series and so the compound contains a C=C double bond.

  A is therefore given the name pent-1-ene and **B** is given the name pent-2-ene.

Naming branched-chain alkenes

To name branched-chain alkenes we need to follow an internationally agreed set of rules again. Look at the compound shown below.

Full structural formula

Shortened structural formulae

$CH_3CH(CH_3)CHCH_2$

- Pick out the longest continuous chain of carbons containing the double bond. This forms the basis for the name. In the compound above the longest continuous chain containing the C=C double bond has **four** carbon atoms in it. This is highlighted in the structure opposite. The prefix for this alkene will be 'but-'.

- Number the chain starting at the end nearest to the C=C double bond. In this example this means that numbering will start at the right-hand end of the chain.

- The first carbon atom of the C=C is carbon-1 and so the name is based on **but-1-ene**.

- The position and name of the branch is now added to the name. In this example there is a methyl branch on carbon-3.

- The systematic name for this branched-chain alkene is 3-methylbut-1-ene.

One use of branched-chain alkenes is to make plastics.

ONLINE TEST

Try the test on naming straight-chain alkenes from full and shortened structural formulae for alkenes up to octane: www.brightredbooks.net/N5Chemistry.

DON'T FORGET

If we consider this pentene molecule and use the above naming rules we arrive at the name pent-2-ene. It is the same as molecule **B**.

DON'T FORGET

It is necessary to identify the longest continuous chain of carbon atoms containing the double bond. Sometimes there are longer continuous chains of carbons in a molecule but the name must be based on the longest chain containing the double bond.

ONLINE

Have a go at the activity on naming branched-chain alkenes from full and structural formulae: www.brightredbooks.net/N5Chemistry.

THINGS TO DO AND THINK ABOUT

1 For the alkenes A–E write:

  (a) the full structural formula

  (b) the shortened structural formula

  (c) the molecular formula.

  A hex-3-ene

  B 2-methylbut-2-ene

  C 4-ethylhex-2-ene

  D 4,5-dimethylhex-3-ene

  E 3-ethylpent-2-ene

2 Name the following alkanes

  (a)

  (b)

  (c) $CH_3CH_2CH(CH_2CH_3)CHCH_2$

HOMOLOGOUS SERIES: MORE HYDROCARBONS

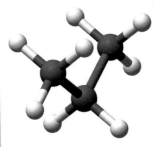

DON'T FORGET

The first member of the cycloalkane family contains three carbon atoms – this is the smallest number of carbons that can be arranged in a cyclic structure.

ONLINE

Try naming cycloalkanes of up to eight carbons from all three formula types at www.brightredbooks.net/N5Chemistry.

ONLINE TEST

Take the test on using general formulae: www.brightredbooks.net/N5Chemistry.

CYCLOALKANES

The **cycloalkanes** are another family of hydrocarbons. Many cycloalkanes are used in motor fuel, kerosene, diesel and other heavy oils. They are also commonly used as **solvents** in industrial processes. They differ from the alkane family in that the carbon atoms are joined in a ring, as the name suggests. They are saturated molecules as all the carbon atoms are joined by single bonds. Like alkanes and alkenes, cycloalkanes are insoluble in water.

The cycloalkane family

The first three members of the cycloalkane family are shown in the table below.

Cycloalkane	Full structural formula	Shortened structural formula	Molecular formula
cyclopropane			C_3H_6
cyclobutane			C_4H_8
cyclopentane			C_5H_{10}

GENERAL FORMULAE

To make it easier to work out the formula of a hydrocarbon, a **general formula** is used. This is a formula that can be applied to a family of hydrocarbons. Take a look at the table below. It contains the molecular formulae for the first four members of the alkane family.

Alkane	Molecular formula
methane	CH_4
ethane	C_2H_6
propane	C_3H_8
butane	C_4H_{10}

The **general formula** for the **alkane** family is C_nH_{2n+2}, where n represents the number of carbon atoms in the compound.

Consider an alkane with ten carbon atoms in it. Based on the general formula this would mean it would have the formula $C_{10}H_{(2 \times 10)+2} \rightarrow C_{10}H_{22}$.

If we have a look at the molecular formulae of the first four members of the alkene family and the cycloalkane family we can work out a general formula for these families as well.

Alkene	Molecular formula
ethene	C_2H_4
propene	C_3H_6
butene	C_4H_8
pentene	C_5H_{10}

Cycloalkane	Molecular formula
cyclopropane	C_3H_6
cyclobutane	C_4H_8
cyclopentane	C_5H_{10}
cyclohexane	C_6H_{12}

Notice that the general formula for the alkene family (C_nH_{2n}) is the same as that for the cycloalkane family (C_nH_{2n}).

contd

The name of a hydrocarbon can allow us to work out its molecular and structural formula. We will use **heptane** as an example. The prefix **hept-** tells us there are **seven** carbon atoms in the molecule. The ending -ane tells us that the molecule belongs to the alk**ane** family and so all the carbon atoms are joined with single bonds. Using the general formula for the alkanes, we can work out that the molecular formula is C_7H_{16}, and from this a structural formula can be written.

Similarly for alkene molecules, the prefix will tell us how many carbon atoms the molecule contains and the -**ene** ending tells us it is a member of the alk**ene** family and that there is a carbon-to-carbon double bond in the molecule. For example in **butene** the prefix **but-** tells us there are four carbon atoms and the -**ene** ending tells us there is a carbon-to-carbon double bond. Using the general formula we can work out that the molecular formula is C_4H_8.

The names of cycloalkane molecules are all prefixed with **cyclo-** and again the -**ane** ending tells us that the carbon atoms are all joined with single bonds. Cyclopentane is from the cycloalkane family and so follows the general formula C_nH_{2n} – the molecular formula for cyclopentane is C_5H_{10}.

We can also use the formula to work out the name of the compound. If a compound has a formula with two fewer hydrogen atoms than an alkane we know it must either contain a carbon-to-carbon double bond or it must contain a ring of carbon atoms.

ONLINE

For more practice on naming cycloalkanes, visit www.brightredbooks.net/N5Chemistry.

ONLINE TEST

Take the 'More Hydrocarbons' test at www.brightredbooks.net/N5Chemistry.

HOMOLOGOUS SERIES

A **homologous series** is a family of compounds that all have the same general formula and have similar chemical properties (react in a similar way). The alkane, alkene and cycloalkane families form three different homologous series. The reactions that these compounds take part in will be looked at later in this chapter.

THINGS TO DO AND THINK ABOUT

As well as having similar chemical properties, the physical properties of a homologous series follow a pattern. Have a look at the information in the tables below.

Alkane series		Alkene series		Cycloalkane series	
alkane	*boiling point (°c)*	*alkene*	*boiling point (°c)*	*cycloalkane*	*boiling point (°c)*
methane	−164	–	–	–	–
ethane	−89	ethene	−104	–	–
propane	−42	propene	−47	cyclopropane	−33
butane	−1	butene	−6	cyclobutane	12
pentane	36	pentene	30	cyclopentane	49

If we look at the alkane homologous series we can see that the boiling points of the alkanes increase (become less negative) as the number of carbon atoms in the molecules increases. This is also true for the melting points. A similar pattern can be seen for the alkene and cycloalkane homologous series.

We can explain this by considering the bonding in a substance. The atoms in an alkane molecule are held together by strong covalent bonds and there are weak interactions between the molecules known as intermolecular forces of attraction. When an alkane is heated, it is the weak intermolecular forces of attraction between the molecules that are broken. The more atoms in the molecule, the stronger these intermolecular attractions are and so more energy and a therefore a higher temperature are needed to break them.

As an analogy, consider a pile of small elastic bands and a pile of large elastic bands. The small elastic bands will be easier to separate than the large ones.

HOMOLOGOUS SERIES: ISOMERS

Did you notice that the general formula for the alkenes is the same as that for the cycloalkanes? This makes alkenes and cycloalkanes **isomers** of each other.

WHAT ARE ISOMERS?

Isomers are defined as compounds with the same molecular formula but different structural formulae.

Look at the structural formulae of butene and cyclobutane:

but-1-ene   cyclobutane

Both molecules have four carbon atoms and eight hydrogen atoms and so have the molecular formula C_4H_8. The atoms are arranged differently and so their structural formulae are different. These two compounds have different physical properties: but-1-ene has a boiling point of –6°C and cyclobutane has a boiling point of 12°C.

ONLINE

For more information on isomers check out www.brightredbooks.net/N5Chemistry.

ALKANE ISOMERS

Have a look at the two molecules below:

pentane   2-methylbutane

Both molecules have the molecular formula C_5H_{12} yet their structural formulae are different. In pentane all five carbon atoms are in a continuous chain whereas in 2-methylbutane four carbon atoms are in a continuous chain, whilst the fifth carbon atom forms a methyl branch on the second carbon atom. The physical properties of these two compounds are different too – pentane has a boiling point of 36°C whereas the boiling point of 2-methylbutane is 30°C. These two molecules are **isomers** of each other.

DON'T FORGET

You will also need to be able to name alkane isomers with up to eight carbons in their structure. See if you can draw and name all the isomers of hexane.

ONLINE TEST

For a test on alkane isomers: www.brightredbooks.net/N5Chemistry.

Have a look at another structural formula for pentane:

Structural formula of pentane

Do you think this is also an isomer of pentane? The answer is **no**. The five carbon atoms are still in a continuous chain. This is pentane, not an isomer of pentane. It has the same structural formula. To understand this it is necessary to think back to the three-dimensional shape of alkane molecules.

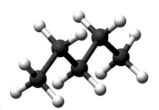

A three-dimensional structural formula for pentane

The structural formula is a two-dimensional version of this and so is commonly drawn as a straight line but so long as all of the carbons are still in a continuous chain then the molecule is still pentane.

ALKENE ISOMERS

There are three main groups of isomers to consider for alkene molecules. The first is based on the position of the double bond in the chain.

Look at the two isomers of pentene below:

pent-1-ene

pent-2-ene

DON'T FORGET

The physical properties of pent-1-ene and pent-2-ene are different. Pent-1-ene has a boiling point of 30°C and the boiling point of pent-2-ene is 36°C.

Both of these isomers have the molecular formula C_5H_{10} and their structures differ only in the position of the double bond. This changes their structural formulae.

The second group of isomers of alkenes is based on branched-chain alkenes. When working out formulae, we should treat this group in a similar way to the branched-chain alkane isomers we met earlier on the previous page.

Look at the two molecules shown below:

pent-2-ene

2-methylbut-2-ene

Pent-2-ene and 2-methylbut-2-ene both have the molecular formula C_5H_{10}. Their structural formulae are different. Pent-2-ene has all five carbon atoms in a continuous chain whereas 2-methylbut-2-ene has only four carbon atoms in a continuous chain and the fifth carbon atom is a methyl branch on the second carbon in the chain.

The third group of alkene isomers were shown in the example at the beginning of this topic using butene and cyclobutane. Alkenes and cycloalkanes have the same general formula and this means that an alkene with a given number of carbon atoms will have an isomer that is a cycloalkane with the same number of carbon atoms.

 THINGS TO DO AND THINK ABOUT

1   Draw the full structural formula of an isomer of butane.

2   There are five possible isomers of C_4H_8.

    (a) Draw the full structural formulae of three alkene isomers.

    (b) Draw the full structural formulae of two cycloalkane isomers.

3   Name the three alkene isomers of C_4H_8 you drew in **2(a)**.

4   Look at the pairs of molecules below. For each pair decide if they are isomers and explain your answer.

    (a)   **(b)**

HOMOLOGOUS SERIES: HYDROCARBON REACTIONS

Alkanes, alkenes and cyclokanes undergo **combustion** reactions, refer to p64 for more. This section looks at other reactions of alkenes. Alkenes contain the carbon-to-carbon double bond functional group and this means they have different chemical properties from alkane compounds.

DON'T FORGET

Saturated hydrocarbons have carbon-to-carbon single bonds like alkanes. Alkenes are unsaturated hydrocarbons as they contain a carbon-to-carbon double bond.

VIDEO LINK

Have a look at the clip 'Testing for unsaturated hydrocarbons' at www. brightredbooks.net/ N5Chemistry.

ONLINE TEST

Check how well you've understood this topic by taking the 'Hydrocarbon reactions' test at www. brightredbooks.net/ N5Chemistry.

TESTING FOR UNSATURATION

If you have carried out a cracking reaction in class you probably tested the products for unsaturation. To do this a few drops of bromine water (bromine solution) are added to the products from the cracking reaction. The yellow-orange bromine solution turns **colourless rapidly** when an unsaturated compound, such as an alkene, is present. This reaction does not take place if an alkane molecule is used. The bromine water does not react with alkanes and so the bromine water remains a yellow-orange colour.

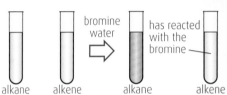

Compound	Reaction with bromine
Saturated	Remains yellow-orange
Unsaturated	Rapidly decolourises

EXAMPLE

If bromine solution is added to an unsaturated compound such as heptene, the following reaction takes place:

$$
\underset{\text{colourless}}{
\begin{array}{c}
\text{H} \; \text{H} \; \text{H} \; \text{H} \; \text{H} \\
| \quad | \quad | \quad | \quad | \\
\text{H}-\text{C}-\text{C}-\text{C}-\text{C}-\text{C}=\text{C}-\text{H} \\
| \quad | \quad | \quad | \quad | \quad | \quad | \\
\text{H} \; \text{H} \; \text{H} \; \text{H} \; \text{H} \; \text{H}
\end{array}}
\; + \; \underset{\text{orange-brown}}{\text{Br} - \text{Br}} \longrightarrow \;
\underset{\text{colourless}}{
\begin{array}{c}
\text{H} \; \text{H} \; \text{H} \; \text{H} \; \text{H} \; \text{Br} \; \text{Br} \\
| \quad | \quad | \quad | \quad | \quad | \quad | \\
\text{H}-\text{C}-\text{C}-\text{C}-\text{C}-\text{C}-\text{C}-\text{C}-\text{H} \\
| \quad | \quad | \quad | \quad | \quad | \quad | \\
\text{H} \; \text{H} \; \text{H} \; \text{H} \; \text{H} \; \text{H} \; \text{H}
\end{array}}
$$

As the bromine water is added to heptene a reaction takes place. One of the atoms from the diatomic bromine molecule adds to one of the carbon atoms involved in the C=C double bond. The second bromine atom adds to the second carbon atom of the C=C double bond. All of the bromine water has reacted and so no yellow-orange colour remains. At the end of the reaction, the mixture is colourless. Notice also that the product contains no carbon-to-carbon double bonds and so the product is saturated. This type of reaction is known as an **addition reaction**. This is a reaction in which a small molecule adds across a carbon-to-carbon double bond, resulting in a saturated product. Alkane molecules do not take part in addition reactions and so bromine water will remain orange when added to an alkane.

ADDITION REACTIONS

Addition of bromine solution to an alkene molecule is one example of an addition reaction. There are some other important addition reactions.

Addition of other halogens

Other group 7 (halogen) molecules can also react with unsaturated molecules. Take a look at the reaction of propene with chlorine:

$$
\begin{array}{c}
\text{H} \\
| \\
\text{H}-\text{C}-\text{C}=\text{C}-\text{H} \\
| \quad | \quad | \\
\text{H} \; \text{H} \; \text{H}
\end{array}
\; + \; \text{Cl} - \text{Cl} \longrightarrow \;
\begin{array}{c}
\text{H} \; \text{Cl} \; \text{Cl} \\
| \quad | \quad | \\
\text{H}-\text{C}-\text{C}-\text{C}-\text{H} \\
| \quad | \quad | \\
\text{H} \; \text{H} \; \text{H}
\end{array}
$$

The products formed when halogens undergo addition reactions with alkenes are called 'dihaloalkanes'. These are saturated molecules and so will have the ending **-ane**.

contd

Addition of water: hydration

The atoms in a water molecule can add across a C=C double bond.

Consider the addition of water to ethene:

$$\underset{\text{ethene}}{\underset{\begin{matrix}H \quad H\\ \diagdown \quad \diagup \\ C=C \\ \diagup \quad \diagdown \\ H \quad H\end{matrix}}{}} \quad + \quad \underset{\text{water}}{H\text{-}O\text{-}H} \quad \longrightarrow \quad \underset{\text{ethanol}}{\underset{\begin{matrix}H \quad OH\\ | \quad | \\ H\text{-}C\text{-}C\text{-}H \\ | \quad | \\ H \quad H\end{matrix}}{}}$$

In a similar way to the addition of bromine, adding water to ethene involves one part of the water molecule (OH) adding to one of the carbon atoms of the double bond and the other part of the water molecule (H) adding to the second carbon atom of the double bond. Again, the product of the reaction is a saturated molecule. This reaction is an important one in the industrial production of an important alcohol – ethanol. You will find out more about alcohols and their uses later.

Addition of hydrogen: hydrogenation

Adding hydrogen across a carbon-to-carbon double bond is called **hydrogenation**. This is an important reaction in industry for the production of many important chemicals, such as pharmaceuticals.

Similarly to addition of bromine, adding hydrogen across a double bond involves one of the atoms of the hydrogen molecule adding to one of the carbon atoms of the double bond and the other hydrogen atom adds to the second carbon atom of the double bond. We can see this by looking at the reaction of propene with hydrogen:

$$\underset{\begin{matrix}H\\ |\\ H\text{-}C\text{-}C=C\text{-}H\\ | \quad | \quad |\\ H \quad H \quad H\end{matrix}}{} \quad + \quad H\text{-}H \longrightarrow \quad \underset{\begin{matrix}H \quad H \quad H\\ | \quad | \quad |\\ H\text{-}C\text{-}C\text{-}C\text{-}H\\ | \quad | \quad |\\ H \quad H \quad H\end{matrix}}{}$$

DON'T FORGET

The product of this reaction is saturated. Can you name this hydrocarbon? What ending does the name have? What does this tell you about the bonds in the molecule?

Addition of hydrogen halides

Another important addition reaction is addition of hydrogen halides. Hydrogen halides are diatomic molecules with a hydrogen atom and a halogen atom. An example of addition of a hydrogen halide is:

$$\underset{\text{but-1-ene}}{\underset{\begin{matrix}H \quad H\\ | \quad |\\ H\text{-}C\text{-}C\text{-}C=C\text{-}H\\ | \quad | \quad | \quad |\\ H \quad H \quad H \quad H\end{matrix}}{}} \quad + \quad \underset{\text{hydrogen chloride}}{H\text{-}Cl} \longrightarrow \quad \underset{\text{2-chlorobutane}}{\underset{\begin{matrix}H \quad H \quad Cl \quad H\\ | \quad | \quad | \quad |\\ H\text{-}C\text{-}C\text{-}C\text{-}C\text{-}H\\ | \quad | \quad | \quad |\\ H \quad H \quad H \quad H\end{matrix}}{}}$$

The product of the reaction is 2-chlorobutane but can you work out another product that will be formed? If the chlorine atom adds onto the **first** carbon atom, the product is 1-chlorobutane. Both 1-chlorobutane and 2-chlorobutane will form during this reaction.

Alkane molecules that have a hydrogen atom replaced by another atom are known as substituted alkanes. Substituted alkanes with a single halogen atom in their structure are incredibly useful molecules as they are very reactive and so can be used to make other products.

DON'T FORGET

As well as being able to write equations for addition reactions using full structural formulae as the examples on these pages have shown, you also need to be able to use shortened and molecular formulae for these equations. For example:
$CH_2\text{-}CH_2 + HCl \rightarrow CH_2Cl\text{-}CH_3$
and $C_4H_8 + HCl \rightarrow C_4H_9Cl$

THINGS TO DO AND THINK ABOUT

1   For each of the following molecules state whether there will be a positive test with bromine water (decolourises).

   (a) C_5H_{12} **(b)** C_2H_6 **(c)** Propene   **(d)** Cyclohexane

2   Two compounds, both with the molecular formula C_6H_{12}, are tested with bromine water. With one compound the bromine water rapidly decolourises, whereas with the other compound there is no reaction and the bromine water remains a yellow-orange colour. Explain these results.

EVERYDAY CONSUMER PRODUCTS

THE IMPORTANCE OF HYDROCARBONS

Hydrocarbon compounds and substituted hydrocarbon compounds are involved in many aspects of our lives.

If we look at a typical morning routine we can see the range and variety of carbon-containing compounds that impact on our lives.

Before you are even awake you will be interacting with carbon-based products. The bed linen you are lying on is likely to be made from a thread that is hydrocarbon based, the alarm clock that wakes you up is likely to contain plastic – plastic materials are based on hydrocarbon compounds. Read more about the important role plastics play in the Chemistry in Society chapter (page 82). The shampoo or shower gel that you use in the shower will contain hydrocarbon-based compounds.

If you have a look around your house you will come across many examples of hydrocarbon-based consumer products. As well as shampoos and other soaps, other personal care products that contain hydrocarbon-based compounds include conditioners, deodorants and cosmetics.

OTHER IMPORTANT COMPOUNDS

There are many compounds containing carbon and hydrogen that are used in our everyday lives. In many ingredients lists there are alcohol compounds and acid compounds listed. These are, again, large families containing many different compounds and their structure and properties will be looked at in more detail later on.

Alcohols

The **alcohol** that most people are aware of is ethanol, which is the alcohol present in alcoholic drinks.

As well as drinks, alcohol compounds are used in a wide variety of commercial products. Benzyl alcohol, for example, is added to many products as it can add fragrance and also acts as a solvent and a preservative. Alcohols are used in skin toners as they can help to dissolve oils.

Alcohols are also good at killing bacteria and so are now commonly used in disinfectant wipes and hand gels.

Carboxylic acids

One of the best-known members of the **carboxylic acid** family is ethanoic acid. One of its most common uses is in vinegar. Carboxylic acids, like alcohols, are also used as preservatives.

Citric acid is another carboxylic acid. It occurs naturally in large quantities in lemons, oranges and limes. It is also listed in the ingredients of shampoo and nail varnish. It is added as a preservative but can also be added to help adjust the pH of the product.

ONLINE TEST

What do you know about the chemistry of everyday consumer products? Test yourself online at www.brightredbooks.net/N5Chemistry.

WHAT ARE ALCOHOL MOLECULES?

Alcohols are a group of compounds based on hydrocarbons that all have a hydrogen atom replaced with the –OH group (**hydroxyl**). We have already seen that they have many uses in our everyday life including alcoholic drinks and as disinfectants.

The alcohol in drinks is called ethanol. Its name does not end in **-ane**. This is because there is an OH group in the molecule, which means that it is not part of the alkane homologous series. The **-an-** in the middle of the name tells us that there are only carbon-to-carbon single bonds in the molecule.

contd

STRAIGHT-CHAIN ALCOHOLS

Let us look at the first four members of the straight-chain alcohols family.

Alcohol	Full structural Formula	Shortened structural formula	Molecular formula
methanol	$H-\overset{\overset{\displaystyle H}{\mid}}{\underset{\underset{\displaystyle H}{\mid}}{C}}-O-H$	CH_3OH or $CH_3{-}OH$	CH_3OH
ethanol	$H-\overset{\overset{\displaystyle H}{\mid}}{\underset{\underset{\displaystyle H}{\mid}}{C}}-\overset{\overset{\displaystyle H}{\mid}}{\underset{\underset{\displaystyle H}{\mid}}{C}}-O-H$	CH_3CH_2OH or $CH_3{-}CH_2{-}OH$	C_2H_5OH
propanol	$H-\overset{\overset{\displaystyle H}{\mid}}{\underset{\underset{\displaystyle H}{\mid}}{C}}-\overset{\overset{\displaystyle H}{\mid}}{\underset{\underset{\displaystyle H}{\mid}}{C}}-\overset{\overset{\displaystyle H}{\mid}}{\underset{\underset{\displaystyle H}{\mid}}{C}}-O-H$	$CH_3CH_2CH_2OH$ or $CH_3{-}CH_2{-}CH_2{-}OH$	C_3H_7OH
butanol	$H-\overset{\overset{\displaystyle H}{\mid}}{\underset{\underset{\displaystyle H}{\mid}}{C}}-\overset{\overset{\displaystyle H}{\mid}}{\underset{\underset{\displaystyle H}{\mid}}{C}}-\overset{\overset{\displaystyle H}{\mid}}{\underset{\underset{\displaystyle H}{\mid}}{C}}-\overset{\overset{\displaystyle H}{\mid}}{\underset{\underset{\displaystyle H}{\mid}}{C}}-O-H$	$CH_3CH_2CH_2CH_2OH$ or $CH_3{-}CH_2{-}CH_2{-}CH_2{-}OH$	C_4H_9OH

The **-ol** ending is used to identify the presence of the –OH functional group, which characterises the alcohols.

The **general formula** for the straight-chain alcohols is $C_nH_{2n+1}OH$.

You may have realised that for the alcohols propanol and butanol there is a choice of carbon atoms to the hydroxyl (OH) group is attached.

Let us look at two straight-chain isomers of pentanol.

The position of the hydroxyl group is denoted using the number and as with naming branched chain alkanes and alkenes, the carbon atoms are numbered in such a way as to give the carbon atom with the hydroxyl group the lowest number.

pentan-1-ol

pentan-2-ol

PROPERTIES OF ALCOHOLS

The smaller, straight-chain alcohols are miscible (soluble) in water. Methanol, ethanol and propanol are completely miscible in water. As the carbon chain of the alcohol increases the solubility of the alcohol decreases. Similar to alkanes, alkenes and cycloalkanes, as alcohols increase in size, their melting and boiling points also increase due to the increasing strength of the intermolecular forces.

Alcohols are used as solvents. Their presence in skincare products such as facial washes and toners is in part due to their ability to dissolve the oils present on the skin.

Alcohol compounds are highly flammable and so can be used as fuels. They have the added advantage that they burn with a much cleaner, less sooty flame than hydrocarbon fuels. A common example that you may have come across is in some camping stoves. Ethanol, made from fermenting renewable plants, is also becoming used more widely as a fuel for vehicles.

 THINGS TO DO AND THINK ABOUT

1 Name the following alcohols:
   (a) $H-\overset{\overset{\displaystyle H}{\mid}}{\underset{\underset{\displaystyle H}{\mid}}{C}}-\overset{\overset{\displaystyle H}{\mid}}{\underset{\underset{\displaystyle H}{\mid}}{C}}-\overset{\overset{\displaystyle H}{\mid}}{\underset{\underset{\displaystyle H}{\mid}}{C}}-\overset{\overset{\displaystyle H}{\mid}}{\underset{\underset{\displaystyle H}{\mid}}{C}}-\overset{\overset{\displaystyle H}{\mid}}{\underset{\underset{\displaystyle H}{\mid}}{C}}-O-H$
   (b) $CH_3CH_2CH_2CH_2CH_2CH_2OH$ (c) $C_7H_{15}OH$

2 Draw full and shortened structural formulae for: (a) octanol   (b) $C_6H_{13}OH$

3 For the examples below, write a shortened structural formula and name the compound:
   (a) $H-\overset{\overset{\displaystyle H}{\mid}}{\underset{\underset{\displaystyle H}{\mid}}{C}}-\overset{\overset{\displaystyle H}{\mid}}{\underset{\underset{\displaystyle H}{\mid}}{C}}-\overset{\overset{\displaystyle H}{\mid}}{\underset{\underset{\displaystyle H}{\mid}}{C}}-\overset{\overset{\displaystyle H}{\mid}}{\underset{\underset{\displaystyle H}{\mid}}{C}}-\overset{\overset{\displaystyle H}{\mid}}{\underset{\underset{\displaystyle H}{\mid}}{C}}-\overset{\overset{\displaystyle H}{\mid}}{\underset{\underset{\displaystyle H}{\mid}}{C}}-\overset{\overset{\displaystyle H}{\mid}}{\underset{\underset{\displaystyle H}{\mid}}{C}}-O-H$
   (b) $H-\overset{\overset{\displaystyle H}{\mid}}{\underset{\underset{\displaystyle H}{\mid}}{C}}-\overset{\overset{\displaystyle H}{\mid}}{\underset{\underset{\displaystyle H}{\mid}}{C}}-\overset{\overset{\displaystyle H}{\mid}}{\underset{\underset{\displaystyle H}{\mid}}{C}}-O-H$

EVERYDAY CONSUMER PRODUCTS: CARBOXYLIC ACIDS

DON'T FORGET

The carboxyl functional group is:

$$\begin{array}{c} O \\ \parallel \\ -C-O-H \end{array}$$

The carboxylic acids are a group of compounds based on hydrocarbons that all have a **carboxyl** functional group.

Vinegar is a solution of a carboxylic acid known as ethanoic acid and the properties of this acid make it useful as a food preservative.

STRAIGHT-CHAIN CARBOXYLIC ACIDS

The table shows the first four members of the straight-chain carboxylic acid family.

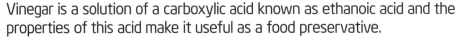

Carboxylic acid	Full structural formula	Shortened structural formula	Molecular formula
methanoic acid		HCOOH	HCOOH
ethanoic acid		CH_3COOH or CH_3-COOH	CH_3COOH
propanoic acid		CH_3CH_2COOH or CH_3-CH_2-COOH	C_2H_5COOH
butanoic acid		$CH_3CH_2CH_2COOH$ or $CH_3-CH_2-CH_2-COOH$	C_3H_7COOH

DON'T FORGET

As with the alkanes and alkenes, the name gives us information about the compound. The prefix tells us how many carbon atoms are in the molecule. The **-an-** tells us that the carbon atoms are all joined by carbon-to-carbon single bonds.

DON'T FORGET

The COOH functional group is known as the **carboxyl** functional group.

The 'oic' ending is used to identify the presence of the COOH functional group which characterises carboxylic acids.

The general formula for the straight-chain carboxylic acids is $C_nH_{2n+1}COOH$.

You will notice that the molecular formula for the straight-chain carboxylic acids is written with the functional group written in full. The true molecular formula for ethanoic acid is $C_2H_4O_2$. There are other compounds (including the ester methyl methanoate) that also have this formula so to make it easier to identify the correct type of carbon compound, the functional group is often used in the molecular formula. This is sometimes called the **functional molecular formula**.

PROPERTIES OF CARBOXYLIC ACIDS

The small carboxylic acid molecules methanoic, ethanoic, propanoic and butanoic acids are miscible (soluble) in water. Larger molecules are not soluble – alkane molecules are insoluble in water so as the size of the alkane chain in a carboxylic acid gets bigger so the solubility decreases. As with other homologous series, the melting and boiling points of the carboxylic acids increases as the size of the molecule increases. This is due to the intermolecular forces of attraction between the molecules becoming stronger.

contd

Carboxylic acids often have unpleasant smells. There are many people who find the smell of ethanoic acid (vinegar) unpleasant. Another example is the smell of 'off' or rancid butter. One of the chemicals responsible for this smell is butanoic acid, which forms when the fat in butter breaks down.

Carboxylic acids behave in a similar way to other types of acid such as dilute hydrochloric acid.

The table below summarises the results of some tests that were carried out in a school lab comparing ethanoic acid ($0.1\,mol\,l^{-1}$) and hydrochloric acid ($0.1\,mol\,l^{-1}$).

Vinegar

Acid	pH	Reaction with magnesium ribbon	Reaction with marble chips
ethanoic acid	3	bubbles of gas	bubbles of gas
hydrochloric acid	1	bubbles of gas	bubbles of gas

The table shows that dilute ethanoic acid solution will react in a similar way to dilute hydrochloric acid solution with magnesium ribbon and marble chips. In fact dilute solutions of carboxylic acids will react with reactive metals, alkalis, metal oxides and metal carbonates to produce salts in a similar way to other dilute acid solutions. The reactions of acids are covered in more detail in the acids and bases sections in the first Unit.

Ethanoic acid is also useful as a cleaning agent, particularly as a limescale remover. The chemical name for limescale is calcium carbonate. As this is a metal carbonate, it reacts with acids to produce a soluble salt, water and carbon dioxide gas.

Limescale

 ONLINE

To revise this topic further, visit the 'Carboxylic Acids' link at www.brightredbooks.net/N5Chemistry.

 DON'T FORGET

Naming salts can be revised in the first chapter – the salts formed from carboxylic acids are shown in the table:

Acid	Salt
Methanoic	Methanoate
Ethanoic	Ethanoate
Propanoic	Propanoate
Butanoic	Butanoate
Pentanoic	Pentanoate
Hexanoic	Hexanoate
Heptanoic	Heptanoate
Octanoic	Octanoate

 THINGS TO DO AND THINK ABOUT

Can you suggest why the same concentrations of dilute ethanoic acid and dilute hydrochloric acid are used in the experiment above?

 ONLINE TEST

Try taking the test on carboxylic acids at www.brightredbooks.net/N5Chemistry.

ENERGY FROM FUELS 1

ENERGY FROM FUELS: AN OVERVIEW

There are many different types of **fuel** that are used in our everyday lives. All of them are burned in oxygen in order to release some of the chemical energy that is stored inside them. This means that burning (or **combustion**) is an **exothermic** reaction.

One of the main uses of hydrocarbons is as a **fuel**. A fuel is a substance that is burned to produce energy.

All hydrocarbons and alcohols undergo **combustion** reactions. Provided there is sufficient oxygen these combustion reactions will produce carbon dioxide and water.

DIFFERENT FUELS

Natural gas

Methane is a common hydrocarbon used as a fuel. Natural gas, which is used in domestic gas supplies, is mainly methane. The balanced equation for the combustion of methane is:

$$CH_4 + 2O_2 \rightarrow CO_2 + 2H_2O$$

Petrol

The energy content of petrol is approximately 48 000 kJ per kg of petrol. This is different from the energy content of other vehicle fuels.

Alcohol

Many modern cars are now able to run on an ethanol/petrol mix. Ethanol can be made by fermentation of plants such as sugar cane and so is a renewable fuel, unlike petrol. Ethanol has an energy content of around 29 700 kJ kg^{-1}.

Methanol is used as a fuel is some racing cars. It is regarded as being safer than petrol. A methanol fuel fire can be extinguished with water, whereas a petrol-based fire cannot. It has an energy content of 20 000 kJ kg^{-1}.

ENERGY CALCULATIONS

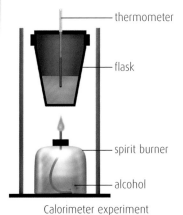

Calorimeter experiment

A calorimeter is used to give energy values for fuels. You may have used equipment like that shown here to carry out an experiment to measure the energy of different fuels.

In a calorimeter, the heat given out by a burning fuel is used to heat a known volume (mass) of water. The change in temperature is measured and this can be used to give an energy value of the fuel if we assume that all of the energy released in burning the fuel has been used to heat the water.

When carrying out the experiment it is necessary to record:

contd

- the mass of water used (the volume can also be measured and, using the conversion 1 kg of water = 1 litre, the mass determined, e.g. 100 cm³ = 0·1 litre = 0·1 kg)

- the initial temperature of the water in the flask

- the final temperature of the water in the flask.

The energy of the fuel can then be calculated using the formula:

$E_h = cm\Delta T$

E_h = energy given out in the reaction, measured in kilojoules (kJ)

c = specific heat capacity of water (this value is given in the data booklet: $4{\cdot}18\,kJ\,kg^{-1}\,°C^{-1}$).

m = mass, in kg, of water being heated

ΔT = rise in temperature of the water in °C

If the same mass of different fuels is burned then the energy content of the different fuels can be compared.

> **EXAMPLE**
>
> The following data was recorded during an experiment in which ethanol was burned using the equipment shown in the diagram on page 66.
>
> volume of water = 100 cm³ (0·1 kg)
>
> initial temperature of the water = 14°C
>
> final temperature of the water = 26°C
>
> From these results we can work out the energy value of the fuel:
>
> 0·1 kg of water was heated
>
> $\Delta T = 26 - 14 = 12°C$
>
> Using the formula, $E_h = cm\Delta T$
>
> $E_h = 4{\cdot}18 \times 0{\cdot}1 \times 12$
>
> $= 5{\cdot}016\,kJ$

ONLINE TEST

Test yourself on energy from fuels at www.brightredbooks.net/N5Chemistry.

THINGS TO DO AND THINK ABOUT

1 The data collected from the combustion of three different fuels is given in the table. Calculate the energy produced by each sample of fuel.

Fuel	Volume of water (cm³)	Initial temperature of water (°C)	Final temperature of water (°C)
1	100	21	29
2	150	19	31
3	200	22	28

2 0·34 g of fuel 2 was burned and 0·56 g of fuel 3 was burned. Suggest which fuel has the highest energy value.

3 It is known that burning fuels derived from fossil fuels, such as crude oil, causes pollution. The carbon dioxide produced in the combustion reaction of hydrocarbons is thought to contribute to **global warming**. Other pollutants are produced too. Any impurities present in petrol will also burn when the petrol burns. One such impurity is sulfur. When sulfur burns it produces a compound that contributes to **acid rain**.

   (a) Write a balanced equation for the combustion of sulfur. Hint: When a substance combusts it combines with oxygen in the air.

   (b) Explain how the burning of sulfur might lead to the formation of acid rain. Try to find out some of the harmful effects of acid rain.

   (c) Write an equation for the combustion of propane.

ENERGY FROM FUELS 2

MORE ENERGY CALCULATIONS

Heating substances other than water

The examples on the previous page all used the energy from a burning fuel to heat up a known quantity (mass) of water. The relationship $E_h = cm\Delta T$ can also be used to calculate the energy involved in heating substances other than water.

E_h = energy and is measured in kilojoules (kJ)

c = specific heat capacity of the substance being heated and is measured in kJkg^{-1}°C^{-1}

m = mass, in kg, of the substance being heated

ΔT = the rise in temperature of the substance in °C

For example, a manufacturer may wish to work out how much energy would be needed to heat an oil filled radiator.

ONLINE

Find the specific heat capacity of a range of different materials on the Digital Zone.

DON'T FORGET

The relationship $E_h = cm\Delta T$ can be found on page 3 of the SQA Data Booklet.

EXAMPLE

Calculate the energy required to heat the oil in a radiator from 15°C to 30°C. The mass of oil in the radiator is 2 kg and the specific heat capacity of the oil is 2·1 kJkg^{-1}°C^{-1}.

Using the relationship, $E_h = cm\Delta T$

specific heat capacity of the oil, c = 2·1 kJkg^{-1}°C^{-1}

mass of substance being heated, m = 2 kg

change in temperature of substance (oil) being heated, ΔT = 15 °C

E_h = 2·1 x 2 x 15

= 63 kJ

SOLVING FOR c, m OR ΔT

As well as calculating the value of E_h, you should be able to calculate values for c, m and ΔT if you are given a value for energy, E_h.

To do this requires the relationship for energy to be rearranged to solve for c, m or ΔT.

$$c = \frac{E_h}{m\Delta T} \qquad m = \frac{E_h}{c\Delta T} \qquad \Delta T = \frac{E_h}{cm}$$

DON'T FORGET

You can rearrange the relationship $E_h = cm\Delta T$ to solve for c, m or ΔT.

EXAMPLE

A male runner generates 401 kJ of heat energy during a run and his body temperature increased by 1·5 °C. The specific heat capacity of the runner is 3·49 kJkg^{-1}°C^{-1}.

Calculate the mass of the runner.

E_h = 401 kJ

c = 3·49 kJkg^{-1}°C^{-1}

ΔT=1·5 °C

Using

$$m = \frac{E_h}{c\Delta T} \qquad\qquad m = \frac{401}{3\cdot49 \times 1\cdot2} \qquad\qquad m = 76\cdot6 \text{ kg}$$

contd

EXAMPLE

When designing a wood burning stove, a manufacturer would need to calculate the maximum temperature the outer, cast iron casing will reach when the wood inside is burning.

If the cast iron casing has a mass of 180 kg and 4500 kJ of energy was produced by the burning wood, what would be the maximum temperature the cast iron casing would reach if the initial temperature was 16·5 °C? The specific heat capacity for cast iron is 0·448 kJkg^{-1}°C^{-1}.

$$\Delta T = \frac{E_h}{cm}$$

$$\Delta T = \frac{4500}{0·448 \times 180}$$

$\Delta T = 55·8\ °C$

ΔT = final temperature – initial temperature

final temperature = ΔT + initial temperature

final temperature = 55·8 + 16·5

final temperature = 72·3 °C

THINGS TO DO AND THINK ABOUT

1. Calculate the energy required to heat a 65 g block of aluminium from 18 °C to 32 °C. The specific heat capacity of aluminium is 0·897 kJkg^{-1}°C^{-1}.

2. A glass mug increased in temperature from 15 °C to 35 °C when heated using 6·7 kJ of heat energy. The mug had a mass of 400 g. Calculate the specific heat capacity of the mug.

3. The best temperature for deep frying potatoes to make into chips is 190 °C. To heat 4·9 kg of oil is heated to this temperature, 1780 kJ energy was used. Calculate the initial temperature of the oil. The specific heat capacity of cooking oil is 2·2 kJkg^{-1}°C^{-1}.

METALS: BONDING AND PROPERTIES

ONLINE

Check out the RSC Visual Elements link at www.brightredbooks.net/N5Chemistry.

METALS IN THE PERIODIC TABLE

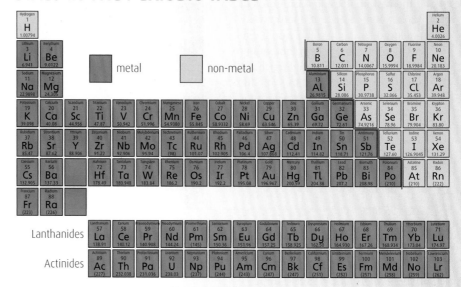

Metals in the Periodic Table

As we learned in the unit on Chemical Changes and Structure, there are 118 chemical elements in the Periodic Table. Of these 118 elements, 94 are metals, which can be found to the left of the dividing step drawn on most Periodic Tables.

METALLIC BONDING AND PROPERTIES

Metal atoms are held together by **metallic bonding**.

A metal atom does not hold on tightly to its outermost electrons and readily forms positive ions. This leaves the outermost electrons delocalised. The term **delocalised** refers to electrons that are not 'attached' to a particular atom. These delocalised electrons are able to move freely throughout the metallic structure or lattice.

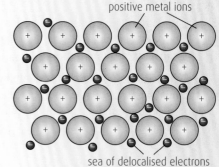

positive metal ions

sea of delocalised electrons

The atoms in a metal are arranged as positively charged metal ions surrounded by a 'sea' of delocalised negatively charged electrons. Metallic bonding is the electrostatic attraction between positive metal ions and the delocalised electrons.

DON'T FORGET

An ion is a charged particle. Metal atoms lose electrons to form positively charged ions.

DON'T FORGET

Metallic bonding is the electrostatic attraction between positive metal ions and the delocalised electrons.

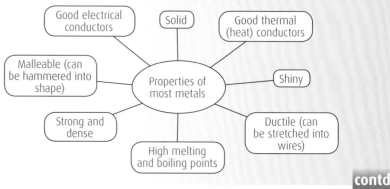

- Good electrical conductors
- Solid
- Good thermal (heat) conductors
- Malleable (can be hammered into shape)
- Properties of most metals
- Shiny
- Strong and dense
- High melting and boiling points
- Ductile (can be stretched into wires)

contd

Metallic bonding explains many of the properties of metals:

- Metallic bonding is strong and it requires lots of energy to break these bonds, which explains the high melting and boiling points of metals.

- The delocalised electrons, which are free to move around, can carry an electric charge, making metals good conductors of electricity.

- Delocalised electrons also explain why metals are such good conductors of heat. Applying heat energy to a metal increases the kinetic energy of these electrons. As they are free to move around, they can transfer this energy to cooler atoms more efficiently than in other materials, which rely solely on the energy being passed between vibrating atoms.

- The malleability and **ductility** of metals is due to the ability of the delocalised electrons to move between the layers of the metal atoms in the metallic structure. Even if a metal changes its shape, the strong metallic bonding is able to remain intact.

USES OF METALS

Metals are very useful materials with a wide variety of uses based on their properties.

- Copper metal, for example, is used to make electrical cables because it can be easily drawn into wires and is a very good conductor of electricity. It is also a good conductor of heat and is used to make saucepans for cooking.

- Gold, silver and platinum are very expensive metals. They are malleable and shiny, making them very desirable for jewellery.

- Aluminium metal is strong, lightweight and doesn't corrode, which makes it a suitable material for building aeroplanes.

- Iron is strong and is used in structures such as fences, bridges and buildings. It is also used in the manufacture of cars, trains and ships. Unfortunately, iron metal, despite its strength, is brittle and will corrode to form rust.

VIDEO LINK

For more, watch the 'Metallic Bonding' video at www.brightredbooks.net/N5Chemistry.

ONLINE TEST

Take the test 'Metals: bonding and properties' at www.brightredbooks.net/N5Chemistry.

THINGS TO DO AND THINK ABOUT

1 The alkali metals are found in group 1 of the Periodic Table. They are soft metals, less dense than water and must be stored under oil because they are so reactive.

   Draw target diagrams for lithium, sodium and potassium. Look closely at these atomic structure diagrams and see if you can explain why the reactivity of these metals increases down the group. (Hint: which part of an atom is involved in bonding?)

2 Aluminium can be used to make window frames whereas iron would be unsuitable for this use because it corrodes when exposed to air and water. Aluminium, however, is a more reactive metal than iron.

   Using the internet, explain why aluminium, despite its reactivity, can be used in this way.

METALS: REACTIONS OF METALS

REACTIVITY SERIES OF METALS

The **reactivity series** of metals places them in order of their reactivity, starting with the most reactive at the top and the least reactive at the bottom. The reactivity series was produced using the results from experiments carried out on a wide variety of metals to investigate their reactivity when reacted in oxygen, water and dilute acids.

The table below shows the order of reactivity and provides a summary of the reactions of metals.

	Metal	Reaction with oxygen	Reaction with water	Reaction with dilute acid
Most reactive ↑	potassium	vigorous reaction	vigorous reaction	violent reaction
	sodium		metal + water ↓	
	lithium			metal + dilute acid ↓
	calcium		metal hydroxide + hydrogen	
	magnesium	metal + oxygen ↓		salt + hydrogen
	aluminium			
	zinc	metal oxide		
	iron		very slow reaction	
	tin			
	lead		no reaction in water	very slow reaction
	copper			
Least reactive ↓	silver	very slow reaction		no reaction with dilute acid
	gold	no reaction in oxygen		

 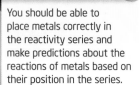
REACTION WITH OXYGEN

Magnesium burns in oxygen with an extremely bright white flame, producing a white powdery solid which is magnesium oxide.

When metals react in oxygen they combine with the oxygen to form a metal oxide:

metal + oxygen → metal oxide

For example, magnesium burns in oxygen to produce magnesium oxide. The word equation for the reaction is:

magnesium + oxygen → magnesium oxide

The balanced chemical equation is:

$2Mg + O_2 \rightarrow 2MgO$

Magnesium oxide is an ionic compound because it is formed when a metal and non-metal bond together.

Ionic equation

We can also write a balanced ionic equation for the reaction taking place that shows ions and state symbols:

$2Mg(s) + O_2(g) \rightarrow 2Mg^{2+}O^{2-}(s)$

REACTION WITH WATER

When metals react in water they produce a metal hydroxide and hydrogen gas.

contd

Looking at the reactivity series, we can see that calcium metal reacts quickly in water whilst magnesium reacts more slowly in water. Metals such as silver and gold found at the bottom of the reactivity series don't react in water at all. This is one of the chemical properties of silver and gold that makes them suitable materials for jewellery and coins.

For example, lithium, an alkali metal, reacts vigorously in water, producing the flammable gas hydrogen and the alkali lithium hydroxide:

lithium + water → lithium hydroxide + hydrogen

$Li(s)$ + $H_2O(l)$ → $LiOH(aq)$ + $H_2(g)$

$Li(s)$ + $H_2O(l)$ → $Li^+(aq) + OH^-(aq)$ + $H_2(g)$

REACTION WITH DILUTE ACID

Metals above copper in the reactivity series react with dilute acids to form a salt and hydrogen gas. For example, zinc metal reacts with dilute hydrochloric acid, producing the salt zinc chloride and hydrogen gas:

zinc + hydrochloric acid → zinc chloride + hydrogen

$Zn(s) + 2HCl(aq)$ → $ZnCl_2(aq)$ + $H_2(g)$

$Zn(s) + 2HCl(aq)$ → $Zn^{2+}(aq) + 2Cl^-$ + $H_2(g)$

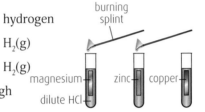

burning splint

magnesium   zinc   copper

dilute HCl

This is not a neutralisation reaction because, although it does produce a salt, it does not produce water.

OXIDATION

The reaction of magnesium metal with oxygen is an example of an **oxidation** reaction. The metal is oxidised to the metal oxide:

$2Mg + O_2 \rightarrow 2MgO$

Oxidation is a term used to describe reactions that involve the loss of electrons. Oxidation occurs whenever a metal forms a compound.

In the reaction above, each of the magnesium atoms loses two electrons to become a positively charged ion.

The ion electron equation for the oxidation of magnesium is:

$Mg \rightarrow Mg^{2+} + 2e^-$

The magnesium atoms must lose electrons to form positively charged ions. These magnesium ions form ionic bonds with the oppositely charged oxygen ions, producing magnesium oxide, which is an ionic compound.

THINGS TO DO AND THINK ABOUT

1   Copper is the metal used to make water pipes. Using the reactivity series explain why copper is a suitable material to use for this purpose.

2   Calcium is a reactive metal but the calcium that is found in your bones, milk and in toothpaste is not reactive. Using your knowledge of chemistry and independent research, explain why.

3   Write a balanced ionic equation for the following reactions:

   (a) copper burning in oxygen to form copper oxide (CuO)

   (b) magnesium reacting with dilute hydrochloric acid (HCL) to form magnesium chloride ($MgCl_2$)

   (c) potassium metal reacting in water to form potassium hydroxide (KOH).

VIDEO LINK

Watch 'Explosive reactions' to see this in action at www.brightredbooks.net/N5Chemistry.

DON'T FORGET

The solution formed when a metal oxide dissolves in water is an alkaline solution.

VIDEO LINK

Watch the demonstration of the reaction between magnesium and dilute hydrochloric acid at www.brightredbooks.net/N5Chemistry.

DON'T FORGET

In the test for hydrogen, the gas burns with a 'pop'.

DON'T FORGET

Please see page 105 for preparation of soluble salt by reacting metal with an acid.

DON'T FORGET

The term oxidation can also be used to describe a reaction in which there is a gain of oxygen.

DON'T FORGET

Oxidation reactions always involve the loss of electrons.

ONLINE TEST

Take the test 'Metals: Reactions of Metals' at www.brightredbooks.net/N5Chemistry.

METALS: EXTRACTION OF METALS

Gold metal trapped in rocks

METAL ORES

Metals such as gold and silver are unreactive and are found uncombined in the Earth's crust as the metallic element.

Most metals are found in the ground combined with other elements in the form of **ionic** compounds called **ores**. An ore is a naturally occurring metal compound found in rocks or underground.

Examples of ores are shown in the table:

Name of ore	Metal compound found in the ore
haematite	iron oxide
bauxite	aluminium oxide
iron pyrite (fool's gold)	iron sulfide

EXTRACTION OF METALS

A metal can be extracted from its ore. Extraction involves separating the metal from the other elements found in the ore. During the extraction, the metal ion is reduced to form the metal atom. The method used to extract a metal depends on the position of the metal in the reactivity series (see page 70).

- Extracting metals from ores is an example of a **reduction** reaction.
- The metal ore is reduced to the metal.

REDUCTION

- Reduction is a term used to describe reactions that involve the gain of electrons.
- Reduction of a metal occurs whenever a metal compound breaks up into its elements.
- The metal in the ore is in the form of a positive ion.
- Metal ions must gain electrons to become metal atoms. For example, when aluminium metal is extracted from its ore, the aluminium Al^{3+} ion in the ore must gain three electrons to form an aluminium atom: $Al^{3+} + 3e^- \rightarrow Al$

METHODS OF EXTRACTION

Using heat alone (silver and gold)

Metals that are unreactive, such as silver and gold, can be found as oxides. They can be extracted easily from the oxide using heat alone. Heat energy is sufficient to break the ionic bonds, allowing the metal to be extracted: $2(Ag^+)_2O^{2-}(s) \xrightarrow{heat} 4Ag(s) + O_2(g)$

Each of the silver ions gains an electron to become a silver atom: $Ag^+ + e^- \rightarrow Ag$

Where do these electrons come from? When extracting metals from their ores the ionic bonds break and the metal and non-metal ions form atoms. Remember earlier in the course, when you learned about ionic bonding:

- Metal atoms usually lose electrons to form positive metal ions.
- Non-metal atoms usually gain electrons to form negative non-metal ions.

When extracting metals from their ores the opposite happens – the ionic bond breaks and the metal and non-metal ions form atoms.

The electrons that the non-metal atom gained from the metal atom in the formation of the ions are returned. The metal ions gain electrons to form metal atoms and the non-metal ions lose electrons to form non-metal atoms.

contd

DON'T FORGET

Unreactive metals are found at the bottom of the reactivity series.

DON'T FORGET

An element is a substance made up of one kind of atom. An element cannot be broken down into anything simpler.

DON'T FORGET

The term reduction can also be used to describe a reaction in which there is a loss of oxygen.

DON'T FORGET

Reactivity series – potassium, sodium, lithium, calcium, magnesium, aluminium, zinc, iron, tin, lead, copper, silver and gold.

DON'T FORGET

Oxygen gas, O_2, is diatomic.

VIDEO LINK

Watch the clip 'Reduction of copper oxide' at www.brightredbooks.net/N5Chemistry.

ONLINE TEST

Take the test 'Metals: extraction of metals' at www.brightredbooks.net/N5Chemistry.

Each oxide ion loses two electrons to form an oxygen atom, which will join with another to form an oxygen molecule: $2O^{2-}(s) \rightarrow O_2(g) + 4e^-$

Heating with carbon or carbon monoxide (copper, lead, tin, iron and zinc)

Heat alone is not sufficient to provide the energy required to extract metals found in the middle of the reactivity series from their ores. Instead, the ores must be heated with a **reducing agent** such as carbon or carbon monoxide. The carbon or carbon monoxide removes the oxygen from the ore and is **oxidised** to carbon dioxide.

The ionic equation for this reaction is: $2Cu^{2+}O^{2-}(s) + C(s) \rightarrow 2Cu(s) + CO_2(g)$

Carbon reduces the copper oxide to copper metal and in the process is oxidised to carbon dioxide gas.

Extracting iron metal

Iron metal is extracted from its ore (haematite, Fe_2O_3) in a blast furnace.

Iron ore, coke (a form of carbon) and limestone (calcium carbonate) are added to the blast furnace. It is called a blast furnace because of the blasts of hot air that are used to increase the temperature inside the furnace and also to supply oxygen.

The carbon (coke) reacts with this oxygen to form, in the first instance, carbon dioxide, which then further reacts with more of the carbon to form carbon monoxide, which is a reducing agent. The carbon monoxide reacts with the iron oxide (ore), reducing it to iron metal and is itself oxidised to carbon dioxide gas.

The high temperatures in the furnace produce the extracted iron in the form of a **molten** liquid.

The ionic equation for this reaction is: $(Fe^{3+})_2(O^{2-})_3(s) + 3CO(g) \rightarrow 2Fe(s) + 3CO_2(g)$

The iron (Fe^{3+}) ion is reduced to an iron atom. The carbon monoxide gains oxygen to become carbon dioxide. A reaction like this is called a **redox** reaction because both reduction and oxidation are taking place.

Electrolysis of molten ore (aluminium, magnesium, calcium, lithium, sodium and potassium)

Very reactive metals are strongly bonded in their ores and cannot be extracted using carbon/carbon monoxide. More energy is required and this is provided by **electrolysis**. Electrolysis is a process that separates an ionic compound into its elements using electrical energy (electricity).

Aluminium metal is extracted from its molten ore (bauxite Al_2O_3) using electrolysis.

Electrolysis requires a d.c. (direct current) supply, which produces a positively charged electrode and a negatively charged electrode. When electricity is passed through the molten ore, the positive ions are attracted to the negative electrode and the negative ions are attracted to the positive electrode.

Reaction at the negative electrode

Aluminium ions gain electrons (reduction) to form the metal atoms, as shown in the following ion electron equation: $Al^{3+}(l) + 3e^- \rightarrow Al(l)$

Reaction at the positive electrode

Oxygen ions lose electrons (oxidation) to form oxygen gas: $2O^{2-}(l) \rightarrow O_2(g) + 4e^-$

ONLINE

Work your way through the electrolysis process online by checking out the 'Electrolysis of aluminium oxide' link at www.brightredbooks.net/N5Chemistry.

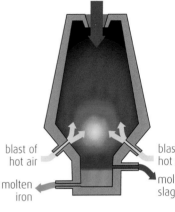

iron ore, coke and limestone

blast of hot air

blast of hot air

molten iron

molten slag

VIDEO LINK

Check out the clip of a blast furnace at www.brightredbooks.net/N5Chemistry.

carbon lining as negative electrodes

carbon positive electrodes

solution of aluminium oxide in molten cryolite

steel tank lined with refractory bricks

molten aluminium collects at the bottom

DON'T FORGET

In solid ionic compounds, ions are arranged in a crystal lattice and are not free to move. When melted (molten) the lattice breaks and the charged ions are free to move around.

THINGS TO DO AND THINK ABOUT

1   Write an ion-detection equation for the reduction Sn^{2+} ions.

2   Cassiterite (SnO_2) is an ore containing tin. Calculate the percentage composition of Cassiterite which is tin.

METALS: ELECTROCHEMISTRY 1

Electrochemistry is a branch of chemistry concerned with chemical reactions that involve electrical currents produced by the transfer of electrons between substances.

ELECTROCHEMISTRY

Electrochemistry involves chemical reactions that take place between different metals to produce a flow of electrons, or electricity.

Electricity can be made by connecting two different metals together in an electrolyte solution to form a simple cell.

The diagrams on the right show a simple electrochemical cell. The chemical reactions taking place are similar to those taking place inside a battery.

Electricity is produced when those two metals are placed in the sodium chloride solution and the electric current flows from the zinc rod to the copper rod through the wires.

A battery is made up of one or more electrochemical cells connected together.

When different pairs of metals are connected together by an electrolyte an electric current is produced because electrons can flow from one metal to the other.

An electrolyte is an ionic compound that can conduct electricity when it is molten (melted) or when dissolved in water to form an aqueous solution for – for example, sodium chloride (common salt) solution.

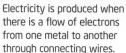

A fruit battery

A fruit battery is a simple electrochemical cell. The juice inside the fruit acts as an electrolyte. Two different metals, for example zinc and copper, are used as electrodes. When the metal electrodes are connected in the cell a chemical reaction takes place.

Zinc is more reactive than copper and gives up some of its electrons, which travel through the wires to the copper electrode producing electricity.

Different pairs of metals produce different voltages.

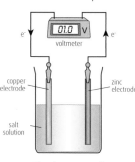

Zinc/copper cell

This difference is used to place metals in their correct order within the **electrochemical series.**

The electrochemical series is a list of both metals and non-metals and places the more reactive metals at the top and the least reactive metals at the bottom.

In an electrochemical cell, electrons travel from a metal higher up in the electrochemical series to a metal lower down in the electrochemical series.

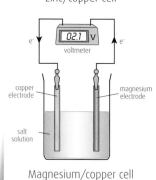

Magnesium/copper cell

Magnesium is higher up the electrochemical series than copper and therefore electrons will travel through the wires from the magnesium metal to the copper metal.

The further apart the metals are from each other in the electrochemical series the greater the voltage produced by the electrochemical cell.

The zinc/copper cell produces a lower voltage than the magnesium/copper cell because magnesium and copper are further apart in the electrochemical series than zinc is from copper.

Displacement reactions

If a metal higher up the electrochemical series is added to an aqueous solution of an ionic compound containing a metal lower down the electrochemical series, then a chemical reaction called **displacement** takes place.

The metal that is higher in the electrochemical series displaces the metal that is lower in the electrochemical series.

DON'T FORGET

Electricity is produced when there is a flow of electrons from one metal to another through connecting wires.

contd

Metal
lithium
potassium
calcium
sodium
magnesium
aluminium
zinc
iron
nickel
tin
lead
hydrogen
copper
silver
mercury
gold

EXAMPLE

In the diagram, the iron (nail) is higher up in the electrochemical series than the copper metal found in the copper sulfate solution. The iron displaces (takes the place of) the copper from its compound, copper sulfate, forming an iron sulfate solution and copper metal.

The word equation for this displacement reaction is:

iron (s) + copper sulfate (aq) → iron sulfate (aq) + copper (s)

If the metal used in this experiment had been lower in the electrochemical series than the copper found in the compound, for example silver, then no displacement reaction would have taken place.

silver (s) + copper sulfate (aq) → no reaction

The results of displacement reactions can be used to place metals correctly in the electrochemical series.

Iron nail — Leave for one week while reaction takes place

blue copper sulfate solution

green iron sulfate solution

Copper metal on iron

Before　**After**

ELECTROCHEMICAL CELLS

Electricity can be produced when half-cells are joined together.

Half-cells consist of a metal electrode placed in a solution of its ions. The metal electrodes are connected by wires. A piece of filter paper soaked in an electrolyte such as sodium chloride solution acts as an **ion bridge** to complete the circuit. An ion bridge completes the circuit by connecting the solutions together, allowing the ions to move from one solution to the other.

Magnesium/copper cell

In the cell shown the electrons flow from the magnesium to the copper through the wires because magnesium is higher than copper in the electrochemical series.

Magnesium/magnesium sulfate half-cell

The magnesium metal loses electrons, which flow through the wires to the copper metal electrode. The half-cell ion electron reaction for this is:

$Mg(s)$　→　$Mg^{2+}(aq)$　+　$2e^-$　oxidation　　metal atoms　　metal ions

Copper/copper sulfate half-cell

Copper ions in the copper sulfate solution pick up the electrons and become copper atoms. The half-cell ion electron equation for this is:

$Cu^{2+}(aq)$　+　$2e^-$　→　$Cu(s)$　reduction metal ions　metal atoms

This is a redox reaction. We can write an equation for the overall reaction taking place by combining the two half-cell reactions and **omitting** the electrons.

Write out the ion electron equations for each of the half-cell reactions:

$Mg(s)$ → $Mg^{2+}(aq)$ + $\cancel{2e^-}$　$Cu^{2+}(aq)$ + $\cancel{2e^-}$ → $Cu(s)$

The number of electrons on each side of the equations is equal, so they cancel each other out.

The overall equation for the redox reaction is: $Mg(s) + Cu^{2+}(aq) \rightarrow Mg^{2+}(aq) + Cu(s)$

Magnesium/copper cell

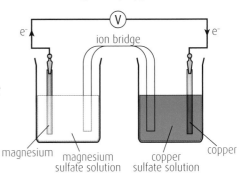

e^-　　Ⓥ　　e^-

ion bridge

magnesium　magnesium sulfate solution　copper sulfate solution　copper

THINGS TO DO AND THINK ABOUT

Everyday objects can be used to make a simple electrochemical cell. A lemon can be used as the electrolyte. A zinc nail and a copper coin are used as the electrodes.

If the copper coin was replaced by an iron nail what would happen to:

(a) the size of voltage produced?　　　　**(b)** the direction of electron flow?

METALS: ELECTROCHEMISTRY 2

ELECTROCHEMICAL CELLS (CONTD)

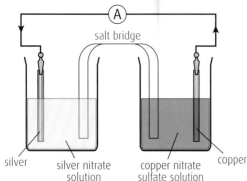

silver — silver nitrate solution — copper nitrate sulfate solution — copper

Copper/silver cell

Copper/silver cell

In this cell the electrons flow from copper to silver because the copper is higher in the electrochemical series than silver.

Copper/copper nitrate half-cell

Copper metal loses electrons, which flow through the wires to the silver metal electrode. The half-cell ion electron reaction for this is:

$$Cu(s) \quad \rightarrow \quad Cu^{2+}(aq) \quad + \quad 2e^- \quad \text{oxidation}$$

metal atoms   metal ions

Silver/silver nitrate half-cell

Silver ions in the silver(I)nitrate solution pick up these electrons to become silver atoms. The half-cell ion electron reaction for this is:

$$Ag^+(aq) \quad + \quad e^- \rightarrow \quad Ag(s) \qquad \text{reduction}$$

metal ions   metal atoms

Writing the equation for the overall reaction:

$$Cu(s) \rightarrow Cu^{2+}(aq) + 2e^-$$

$$2Ag^+(aq) + 2e^- \rightarrow 2Ag(s)$$

The number of electrons on each side of the equations is not equal so before we can combine the equations the number of electrons on each side must be equal.

If we multiply the silver/silver nitrate half-cell equation by 2 then the number of electrons will be the same as those for the copper/copper nitrate half-cell reaction:

$$2Ag^+(aq) + 2e^- \rightarrow 2Ag(s)$$

The two equations can now be combined, omitting the electrons, which have now cancelled each other out. The overall equation for the redox reaction is:

$$Cu(s) + 2Ag^+(aq) \rightarrow Cu^{2+}(aq) + 2Ag(s)$$

ELECTROCHEMICAL CELLS INVOLVING NON-METALS

The electrochemical series is not restricted just to metals, it also includes non-metals. This means that not all electrochemical cells need to contain metals only. In fact non-metals can be used in half-cells if a carbon (graphite) rod is used as the electrode.

As long as you have one substance that is able to give away electrons and another that accepts the electrons, then you can construct a cell. You also need an ion bridge to complete the circuit.

Nickel/iodine cell

The electrons flow from the nickel to the carbon because they flow from the substance that is higher in the electrochemical series, in this case nickel, to the substance lower in the electrochemical series. The further apart the substances are in the series, the higher the voltage produced by the cell.

DON'T FORGET

Graphite is a covalent network form of carbon, which can conduct electricity.

contd

Nickel/nickel sulfate half-cell

The nickel atoms lose electrons and change into nickel(II) ions. The half-cell reaction is:

$Ni(s) \rightarrow Ni^{2+}(aq) + 2e^-$ oxidation

Iodine/iodide half-cell

The iodine molecules pick up these electrons and are changed into iodide ions. The half-cell reaction is:

$I_2(aq) + 2e^- \rightarrow 2I^-(aq)$ reduction

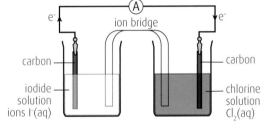

Nickel/iodine cell

Combining the two half-equations gives the following overall equation for the redox reaction:

$Ni \rightarrow Ni^{2+} + 2e^-$

$I_2 + 2e^- \rightarrow 2I^-$

Therefore, the overall redox equation (omitting electrons and including all state symbols) is:

$Ni(s) + I_2(aq) \rightarrow Ni^{2+}(aq) + 2I^-(aq)$

Iodide/chlorine cell

In this cell there are no metals involved and the electrodes are both made from carbon (graphite). The electrons flow from left to right because I^- (aq) is higher in the electrochemical series than $Cl_2(aq)$.

As the electrons are flowing from the left-hand side to the right-hand side, the iodide ions on the left must be losing electrons (oxidation), as shown in the ion-electron equation below:

$2I^-(aq) \rightarrow I_2 + 2e^-$ oxidation

This means that in turn the chlorine molecules will gain these electrons (reduction), as shown in this ion-electron equation:

$Cl_2 + 2e^- \rightarrow 2Cl^-(aq)$ reduction

Combining the two equations produces the overall redox reaction:

$2I^-(aq) + Cl_2 (aq) \rightarrow I_2 (aq) + 2Cl^-(aq)$ redox

> **DON'T FORGET**
>
> Before combining half-cell equations ensure that the numbers of electrons on each side of the equations are the same so they can cancel each other out.

> **ONLINE TEST**
>
> Take the 'Electrochemistry' test at www.brightredbooks. net/N5Chemistry.

THINGS TO DO AND THINK ABOUT

1   Which of the two cells – magnesium/copper (p77) or copper/silver – will produce the greatest voltage? (Hint: You may wish to look at each metal's position in the electrochemical series.)

2   Consider the following electrochemical cell:

   (a) Identify the direction of electron flow

   (b) Write a combined equation for the overall reaction taking place.

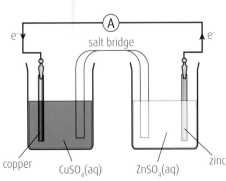

3   An electrochemical cell can be produced using two carbon electrodes, one of which is placed in a glass beaker containing sodium sulfite solution and the other in a glass beaker containing bromine solution. The electrodes are connected with wires and the beakers by an ion bridge.

The sulfite ions are oxidised as shown in the following equation:

$SO_3^{2-}(aq) + H_2O(l) \rightarrow SO_4^{2-}(aq) + 2H^+(aq) + 2e^-$

   (a) Draw a labelled diagram of the electrochemical cell.

   (b) Draw an arrow on your diagram to show the direction of electron flow.

   (c) The bromine solution is reduced. Write an ion-electron equation for this reaction.

PLASTICS, POLYMERS AND POLYMERISATION

DON'T FORGET

Plastics are made up of many repeating units joined together to form long-chained molecules called polymers.

DON'T FORGET

The term monomer means 'one part' (the prefix mono = one).

DON'T FORGET

The name of the polymer is made from putting the prefix 'poly' in front of the systematic name of the monomer, which is enclosed in brackets.

DON'T FORGET

Alkenes are unsaturated hydrocarbon molecules that contain a carbon-to-carbon double covalent bond in their structure.

WHAT IS A PLASTIC (POLYMER)?

Plastics are a group of important synthetic materials called **polymers**.

A polymer is an extremely large molecule made up of single-unit **monomers** joined together to form a long chain of **repeating units**. The term polymer means 'many parts' (poly = many and mer = a part).

The name of the polymer ion can be deduced from the name of the monomer:

Polymers are formed when many smaller monomer units join together in a repeating pattern through a chemical process called polymerisation.

Name of monomer	Name of polymer
ethene	poly(ethene) also known as polythene
vinyl chloride	poly(vinyl chloride) also known as PVC
propene	poly(propene)
styrene	poly(styrene)

Polymerisation of ethene

section of a polythene molecule

ADDITION POLYMERISATION

Addition polymerisation takes place when the monomer involved is an unsaturated molecule, for example an alkene.

Poly(ethene) is an addition polymer formed when many ethene monomers join together:

The ethene monomer molecules join together because the double bonds open up, providing unpaired electrons on either side of the molecules. The unpaired electrons from the adjacent molecules share to make new covalent bonds, which link the monomers together to form part of a single poly(ethene) polymer chain.

Note that the section of polymer no longer contains double bonds as they were broken during the polymerisation process. All addition polymers have a 'backbone' made up entirely of carbon atoms. The end bonds have been left open to indicate that this only represents a small section of a larger polymer chain.

Section of poly(ethene)

contd

Drawing the monomer

Now consider the polymer poly(propene). The diagram to the right represents the structural formula of propene, C_3H_6.

Propene

The diagram below represents three monomers of propene drawn to show the carbon-to-carbon double bond as the centre of each monomer. It is important to draw the monomers in this way because when they join together by addition polymerisation they will join through the opening up of these double bonds.

Poly(propene)

Identifying the repeating unit and the monomer

If the structure of a polymer is given, it is possible to identify both the repeating unit and the monomer from which the polymer was formed. Let us look at poly(vinyl chloride) as an example.

The repeating unit is:

Vinyl chloride repeating unit   A section of PVC

Note: The square brackets are drawn to show that this is one unit, where n = number of repeating units in the polymer chain. Again the end bonds are left open.

Because the monomer for an addition polymer is an unsaturated molecule, it must contain a carbon-to-carbon double bond:

Vinyl chloride monomer

To succeed at National 5, you need to be able to correctly draw a section of a polymer chain from a given monomer, showing at least three monomers joined together, or correctly identify and draw the repeating unit or monomer from a given section of the structure of an addition polymer.

THINGS TO DO AND THINK ABOUT

Poly(butene) is an addition polymer made up of butene, C_4H_8, monomers.

(a) Draw a section of poly(butene) showing three monomer units joined together.

Three butene monomers

(b) Draw the repeating unit for poly(butene).

VIDEO LINK

Check out the 'Addition polymeristation' clip for more at www.brightredbooks.net/N5Chemistry.

ONLINE TEST

Test your knowledge of polymerisation online at www.brightredbooks.net/N5Chemistry.

FERTILISERS: COMMERCIAL PRODUCTION

PLANTS FOR FOOD

All of our food is produced from either animal or plant products.

The world's population is increasing, with current figures exceeding 7 billion. A report produced by the United Nations predicts that by 2050 the population could be as high as 10 billion. With increased population comes increased demand for food.

Farmers have to find ways to produce sufficient crops to meet the increasing demand for plants as food.

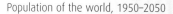

Population of the world, 1950–2050

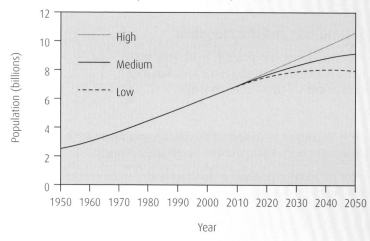

VIDEO LINK

Check out the clip 'What are fertilisers?' at www.brightredbooks.net/N5Chemistry.

WHAT DO PLANTS NEED IN ORDER TO GROW?

Plants need water, sunlight, gases from the air, soil and space to grow. They also need the chemical elements shown in the table.

Plants absorb nutrients through their roots from the soil. These nutrients contain elements and compounds that provide the nitrogen, phosphorus and potassium essential for healthy plant growth. Farming removes these nutrients from the soil when crops are harvested, which means that further plant growth is affected by the lack of nutrients in the soil.

Synthetic fertilisers can be added to the soil to replace these lost nutrients to enable farmers to reuse the land immediately to plant more crops. A fertiliser is a substance that contains the essential elements needed by plants for healthy growth.

Chemical element	Element comes from:
carbon	carbon dioxide in the air
oxygen	water in the air and soil, carbon dioxide in the air
hydrogen	water in the air and soil
nitrogen	nutrients in the soil and gas in the air
phosphorus	nutrients in the soil
potassium	nutrients in the soil

NATURAL FERTILISERS

A natural fertiliser is made from the waste products made by plants and animals, and provides a useful source of the element nitrogen. Plant **compost** and animal **manure** are natural fertilisers.

Bonemeal and fishmeal are made from crushed bones and other non-edible parts of animals and fish that are removed during food production. They can be used as animal food and also as natural fertilisers.

Natural fertilisers are environmentally friendly as they make use of natural waste. There are, however, insufficient quantities for these to be the only source of fertiliser. This is where chemistry can play a part in solving an everyday problem.

DON'T FORGET

Synthetic means man-made.

SYNTHETIC (OR ARTIFICIAL) FERTILISERS

These are artificial fertilisers made through large-scale industrial chemical processing.

Synthetic fertilisers are made from ionic compounds that contain the essential elements nitrogen (N), phosphorus (P) and potassium (K). For this reason they are often referred to as NPK fertilisers.

These ionic compounds are in the form of soluble salts. They dissolve in damp soil, allowing the essential elements to be absorbed through the roots of plants.

Examples of synthetic fertilisers are shown in the table:

Name of fertiliser	Chemical formula	Essential elements (N, P or K)
ammonium nitrate	NH_4NO_3	N
ammonium phosphate	$(NH_4)_3PO_4$	N and P
potassium nitrate	KNO_3	K and N

FERTILISERS: MAKING FERTILISERS 1

DON'T FORGET

Fertilisers must be soluble so they can dissolve in damp soil, enabling plants to absorb the fertiliser compounds through their roots.

DON'T FORGET

The products of an acid and alkali neutralisation reaction are always a salt and water.

DON'T FORGET

Soluble metal hydroxides are alkalis which belong to a group of chemicals called bases.

MAKING FERTILISERS: AN OVERVIEW

Fertilisers are soluble salts. In the Chemical Changes and Structure unit you learned about neutralisation reactions. Fertilisers can be made in the laboratory using a simple neutralisation reaction.

acid + base → salt + water

Potassium nitrate is a salt made from the neutralisation reaction between potassium hydroxide and nitric acid.

Word equation:

nitric acid + potassium hydroxide → potassium nitrate and water

Chemical equation:

$HNO_3(aq) + KOH(aq) \rightarrow KNO_3(aq) + H_2O(l)$

The aqueous solution produced can be dried using evaporation, which removes the water, leaving solid potassium nitrate salt behind.

It makes a good fertiliser because:

- it is soluble and therefore will dissolve in the soil
- it contains the essential elements for plant growth – potassium (K) and nitrogen (N).

COMMERCIAL PRODUCTION OF AMMONIUM NITRATE FERTILISER

In 2013 the world demand for fertilisers was approximately 181 million tonnes, with Europe accounting for around 13% of this demand.

Some 60% of the fertiliser used is nitrogen-based, so the industrial preparation of nitrogen-based fertilisers is a major part of the world's chemical industry. Ammonium nitrate, NH_4NO_3, is commonly used in farming as a fertiliser because of its high nitrogen content.

Making ammonium nitrate

Word equation:

ammonia(aq) + nitric acid(aq) → ammonium nitrate(aq)

Chemical equation:

$NH_3(aq) + HNO_3(aq) \rightarrow NH_4NO_3(aq)$

From the equations you can see that the reactants needed to make ammonium nitrate are the chemicals ammonia and nitric acid.

What is ammonia?

VIDEO LINK

Watch the 'Ammonia fountain explained' video at www.brightredbooks.net/N5Chemistry.

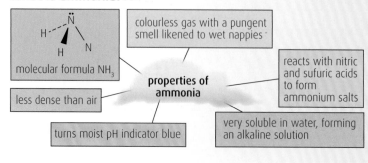

molecular formula NH_3

colourless gas with a pungent smell likened to wet nappies

reacts with nitric and sufuric acids to form ammonium salts

less dense than air

properties of ammonia

turns moist pH indicator blue

very soluble in water, forming an alkaline solution

contd

The fountain experiment demonstrates the solubility of ammonia gas.

The Haber process

The Haber process (also known as Haber–Bosch process) is used for the industrial preparation of ammonia, NH_3.

Fritz Haber was a German chemist who, along with the engineer Carl Bosch, developed the process still used today to produce ammonia from atmospheric nitrogen. In 1918 Haber was awarded the Nobel prize for chemistry in recognition of the importance of his work.

In the Haber process, nitrogen gas is combined with hydrogen gas to form ammonia gas. The nitrogen gas comes from the air and the hydrogen gas comes from methane, CH_4, or water, H_2O.

Word equation:

nitrogen + hydrogen $\rightleftharpoons$ ammonia

Balanced chemical equation:

$$N_2(g) + 3H_2(g) \rightleftharpoons 2NH_3(g)$$

The double arrow indicates that this is a reversible reaction. This means that as some of the nitrogen and hydrogen is combining to form the ammonia, some of the ammonia formed is also breaking back down into nitrogen and hydrogen.

To obtain the best yield of ammonia at the most economic cost the following conditions are used:

- Moderate temperature – at low temperatures the forward reaction would be too slow to be economically viable. If higher temperatures are used, although the rate of reaction would increase so would the backwards reaction resulting in less product.

- An iron catalyst is used to increase the reaction rate.

- High pressure of around 200 atmospheres.

This results in a mixture of nitrogen, hydrogen and ammonia. The mixture is cooled, which turns the ammonia gas into a liquid and allows it to be removed. The nitrogen and hydrogen that remain are not wasted but instead are recycled back into the process. This improves the cost-effectiveness of the reaction.

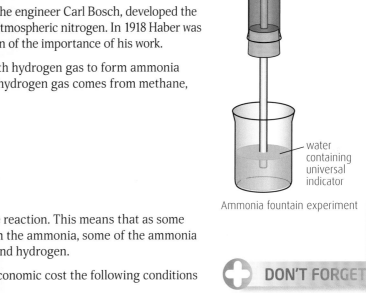

Ammonia fountain experiment

 DON'T FORGET

You will be expected to know the reaction conditions and the name of the catalyst used in the Haber process.

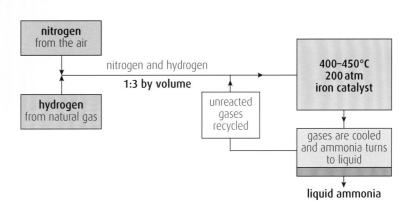

 THINGS TO DO AND THINK ABOUT

1   Potassium sulfate and potassium phosphate are ionic compounds that could be used as fertilisers.

    (a) Suggest two reasons why both of these compounds are suitable for use as fertilisers.

    (b) Write equations for the two reactions that could be carried out in a laboratory to prepare these compounds.

2   The Haber process was developed during the First World War to produce ammonia for use by the German forces in the manufacture of explosives. Use the internet to research the history of the Haber process and the scientists involved.

 VIDEO LINK

Watch the clip 'Formation of ammonia in the Haber Process' at www.brightredbooks.net/N5Chemistry.

ONLINE TEST

Test yourself on 'Making fertilisers' online at www.brightredbooks.net/N5Chemistry.

FERTILISERS: MAKING FERTILISERS 2

Nitric acid, HNO$_3$(aq), is the acid required to neutralise ammonia to form the salt ammonium nitrate, NH$_4$NO$_3$, which is widely used as a fertiliser due to its high nitrogen content.

DON'T FORGET

A catalyst speeds up the reaction without being used up, so the platinum metal can be reused for future reactions.

CATALYTIC OXIDATION OF AMMONIA

Making nitric acid in the laboratory

Nitric acid can be prepared on a small scale using the catalytic oxidation of ammonia. The catalyst used is the metal platinum.

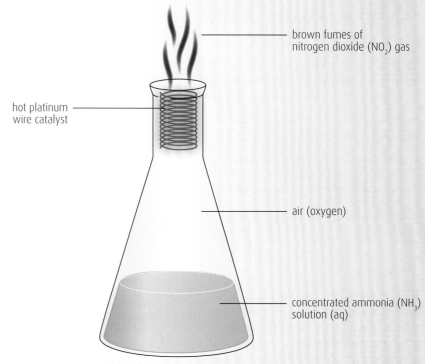

hot platinum wire catalyst

brown fumes of nitrogen dioxide (NO$_2$) gas

air (oxygen)

concentrated ammonia (NH$_3$) solution (aq)

Catalytic oxidation of ammonia

DON'T FORGET

An exothermic reaction releases heat energy to the surroundings.

VIDEO LINK

Check out the 'Catalytic oxidation of ammonia' clip at www.brightredbooks.net/N5Chemistry.

This reaction must be carried out in a fume cupboard because both the concentrated ammonia and nitrogen dioxide are toxic.

The platinum wire is coiled to increase the surface area available for reaction. First, the wire is heated in a Bunsen flame until it is glowing hot. It is then plunged into the flask of ammonia to speed up the reaction. The wire continues to glow red because the reaction is exothermic and therefore the wire remains hot. Fumes of nitrogen dioxide gas are produced, which dissolve in water to form the nitric acid.

INDUSTRIAL PREPARATION OF NITRIC ACID

The Ostwald process

The Ostwald process is used in the chemical industry to produce nitric acid. In 1902 the Latvian chemist Wilhelm Ostwald developed the Ostwald process, which is still used as the industrial method for producing nitric acid. It is based on the catalytic oxidation of ammonia and gained Oswald a Nobel prize in chemistry for his work on catalysed reactions. The two reactants needed for the Ostwald process are ammonia gas and oxygen gas.

contd

Ammonia gas is produced using the Haber process and the oxygen gas comes from the air. A platinum catalyst speeds up the reaction, allows the reaction to take place at lower temperatures and provides a surface on which the reaction takes place. The reaction is carried out at atmospheric pressure and at a temperature of 600–900°C.

A flow diagram of the reaction is shown below:

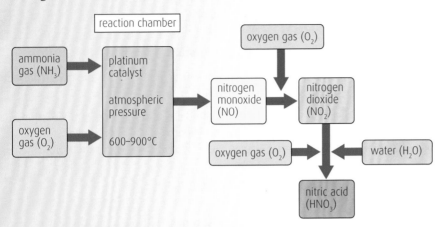

DON'T FORGET

This is an exothermic reaction which generates sufficient heat to keep the platinum catalyst hot.

The ammonia and oxygen gases are passed over thin layers of gauze containing the platinum catalyst. This causes the catalytic oxidation of the ammonia and forms nitrogen monoxide gas and water.

The equations for the reaction taking place are as follows:

Step one – catalytic oxidation of the ammonia:

$2NH_3(g) + 2\tfrac{1}{2}\,O_2(g) \rightarrow 2NO(g) + 3H_2O(g)$

Step two – the nitrogen monoxide reacts in oxygen to form nitrogen dioxide:

$2NO(g) + O_2(g) \rightarrow 2NO_2(g)$

Step three – the nitrogen dioxide reacts with more oxygen and water to form nitric acid:

$4NO_2(g) + O_2(g) + 2H_2O(l) \rightarrow 4HNO_3(aq)$

Ammonia produced in the Haber process and nitric acid produced by the Ostwald process can now be reacted together to form fertilisers.

VIDEO LINK

Check out the 'Catalytic formation of nitrogen dioxide in the Ostwald Process' clip at www.brightredbooks.net/N5Chemistry.

ONLINE TEST

Try taking the 'Making fertilisers 2' test online at www.brightredbooks.net/N5Chemistry.

THINGS TO DO AND THINK ABOUT

1   Write the chemical formula for:

   (a) ammonia

   (b) nitric acid

   (c) nitrogen monoxide

   (d) nitrogen dioxide.

2   Ammonium nitrate and ammonium phosphate are ionic compounds. Write an ionic formula for each of these compounds.

3   Nitrogen dioxide gas formed in nature can dissolve in rainwater to produce nitric acid. Use the internet or other reference sources to learn more about this process and the natural conditions required.

NUCLEAR CHEMISTRY: RADIOACTIVITY AND PROPERTIES

ONLINE

Learn more about background radiation by checking out the link at www.brightredbooks.net/N5Chemistry.

BACKGROUND RADIATION

Background radiation is all around us. It can be natural or artificial.

Most background radiation comes from natural sources such as cosmic rays from outer space or rocks that emit the radioactive gas radon.

Human activity has produced artificial background radiation through the creation and use of artificial sources, such as the radioactive waste from nuclear power stations. The use of X-rays in medicine is the largest contributor to artificial radiation.

DON'T FORGET

Isotopes are atoms of the same element that contain the same number of protons and electrons, but the number of neutrons in each of their nuclei is different.

WHAT IS RADIOACTIVITY?

Most elements exist as a mixture of different **isotopes**. Hydrogen has three isotopes:

Isotopes of hydrogen

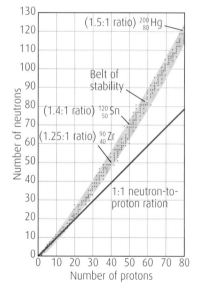

Radioisotopes are radioactive isotopes of elements.

While many isotopes are stable, others are not. The stability of any atom depends on the neutron-to-proton ratio. The lower the atomic number, the more stable the nuclei, as they have approximately equal numbers of neutrons and protons. As the atomic number of an element increases, so does the neutron-to-proton ratio.

When the number of neutrons is plotted on a graph against the number of protons we can produce a band of stability. The lighter nuclei, which have equal numbers of protons and neutrons, are stable but the heavier nuclei, which have an increasing difference in the numbers of neutrons and protons, lie out on this band of stability and are unstable.

Most of the nuclei beyond atomic number 83 are unstable but they can gain stability if they decrease in mass through the loss of protons and neutrons.

Radioactivity is the result of these unstable atomic nuclei decaying. As they decay, they emit particles and energy until the ratio of neutron to protons comes within the range of a stable nucleus.

This **radioactive decay** is completely spontaneous.

TYPES OF RADIATION AND PROPERTIES

In 1896 the French scientist Henri Becquerel discovered the invisible phenomenon we know today as radiation. His experiments, and those of fellow scientists Pierre and Marie Curie, led to the discovery of radioactivity. In 1903 Becquerel and the Curies were awarded a joint Nobel prize for physics in recognition of the importance of their work.

Radiation was found to be affected by a magnetic or electric field. The deflection towards the positive and negative terminals is evidence for different types of radiation.

contd

There are three types of nuclear radiation: alpha (α), beta (β) and gamma (γ) radiation. Alpha and beta radiation are made up of charged particles:

- **Alpha particle**s are helium nuclei and contain two protons and two neutrons.

- **Beta particles** are high-energy electrons ejected from the nucleus. This may seem strange as the nucleus of an atom does not contain any electrons. These high-energy electrons are produced when a neutron breaks up to produce a proton and an electron. The electron is immediately ejected from the nucleus, with high energy, as a beta particle:

$${}_{0}^{1}n \longrightarrow {}_{1}^{1}p + {}_{-1}^{0}e \quad \beta$$

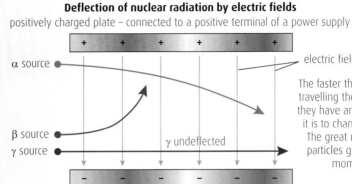

Deflection of nuclear radiation by electric fields

positively charged plate – connected to a positive terminal of a power supply

α source

β source
γ source

γ undeflected

electric field lines

The faster the beta particles are travelling the more momentum they have and the more difficult it is to change their trajectory. The great mass of the alpha particles gives them a great momentum too.

negatively charged plate – connected to a negative terminal of a power supply

Gamma radiation is electromagnetic waves that are emitted from within the nucleus of an atom. It is a similar radiation to that of X-rays but is of higher energy.

These three types of radiation have different properties. The specific properties that are required as knowledge for this course are given in the table below:

Name	Nature	Symbol	Charge	Mass (amu)	Penetrating power
alpha (α)	helium (He) nucleus	${}_{2}^{4}He$ or α	2+	4	low: travels a few cm in air; stopped by a sheet of paper
beta (β)	high-energy electron	${}_{-1}^{0}e$ or β	1–	$\frac{1}{2000}$	medium: travels over a metre in air stopped by thin metal foil
gamma (γ)	waves (EMR)	γ	none	none	high: 10 cm of lead or 0.1 m of concrete

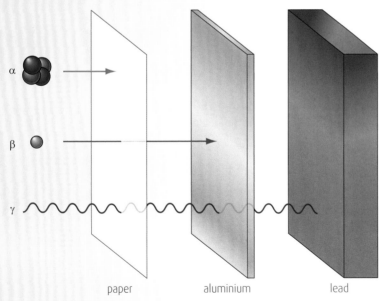

α

β

γ

paper

aluminium

lead

The relative penetrating powers of the three types of radiation

DON'T FORGET

You need to learn the specific properties of the three types of radiation, including mass, charge and ability to penetrate different materials.

ONLINE TEST

Test your knowledge of alpha, beta and gamma radiation online at www.brightredbooks.net/N5Chemistry.

THINGS TO DO AND THINK ABOUT

Torness nuclear power station is located near Dunbar in East Lothian, Scotland. It operates an advanced gas-cooled reactor and uses ${}^{235}U$ as its fuel. Use the internet to research nuclear power. Consider the pros and cons of generating electricity using nuclear power.

NUCLEAR CHEMISTRY: RADIOACTIVE DECAY AND HALF-LIFE

ONLINE

Explore nuclear equations further at www.brightredbooks.net/N5Chemistry.

DON'T FORGET

The atomic number of an atom is equal to the number of protons. The mass number of an atom is equal to the total number of protons and neutrons.

DON'T FORGET

The mass of both protons and neutrons is 1 amu.

DON'T FORGET

You should be able to write nuclear equations for alpha and beta decay for a given radioisotope.

VIDEO LINK

Check out the 'Radioactive decay and half-life' clip at www.brightredbooks.net/N5Chemistry.

RADIOACTIVE DECAY

Nuclear reactions take place as radioisotopes decay. The changes that take place to the nuclei depend upon the type of radiation involved.

Alpha decay

In alpha decay, the nucleus emits an alpha particle. An alpha particle is a helium nucleus, so in effect the nucleus is losing two protons and two neutrons. This means that the atomic number decreases by two and the mass number decreases by four.

Consider the following example: $^{210}_{84}\text{Po} \longrightarrow \, ^{206}_{82}\text{Pb} + \, ^{4}_{2}\text{He}$

The polonium nucleus has 84 protons, making it unstable. It emits an alpha particle, reducing the atomic number by two and the mass number by four, to produce a stable nucleus.

Beta decay

A beta particle is a high-energy electron emitted when a neutron breaks up. The effect on the nucleus is that the atomic number increases by one (one proton) but there is no change to the mass number.

Consider the following example: $^{222}_{88}\text{Ra} \longrightarrow \, ^{222}_{89}\text{Ac} + \, ^{0}_{-1}\text{e}$

The radium nucleus emits a beta particle, increasing the atomic number by one to form actinium. As the actinium nucleus is also radioactive, further decay will take place until a stable nucleus is formed.

Gamma decay

Gamma decay occurs because the nucleus needs to lose energy.

Consider the following example: $^{60}_{27}\text{Co} \longrightarrow \, ^{60}_{27}\text{Co} + \, ^{0}_{0}\gamma$

As gamma rays have no mass or charge, their emissions will have no effect on either the mass number or atomic number of the nuclei.

In a balanced nuclear equation:

- the total mass number on the left-hand side (reactants) is equal to the total mass number on the right-hand side (products)
- the total atomic number on the left-hand side is equal to the total atomic number on the right-hand side.

MEASURING RADIOACTIVITY

When a sample of a radioisotope undergoes decay, the decay of individual nuclei is a completely random event. It is independent of any chemical or physical factor. However, it is possible to measure radioactivity using a Geiger counter. This is actually a Geiger–Müller tube with some form of counter attached, which usually tells us the number of particles detected per minute, given in counts per minute.

A Geiger counter

HALF-LIFE

It is also possible to calculate how much of a radioisotope has decayed and how much remains after a given time period using the **half-life** of the isotope.

- Half-lives can be anything from a matter of seconds to millions of years.

- The half-life of any isotope is independent of temperature, pressure, concentration, chemical state or the mass of the sample being tested.

- The half-life of any radioisotope is the time taken for half the radioactive nuclei to decay and is often abbreviated to $t_{1/2}$.

Consider the graphs, which shows a typical decay curve for sodium-24:

From the decay curve we can see that it takes 15 hours for the mass of the sample/activity of the sample to halve, i.e. for the mass to decay from 100 g to 50 g and for the activity to fall from 400 to 200 counts per minute. Therefore we can see that the half-life of Na-24 is 15 hours.

Different isotopes have different half-lives. For example, iodine-131 has a half-life of 8 days and carbon-14 has a half-life of 5730 years.

ONLINE TEST

Test yourself on this topic at www.brightredbooks.net/N5Chemistry.

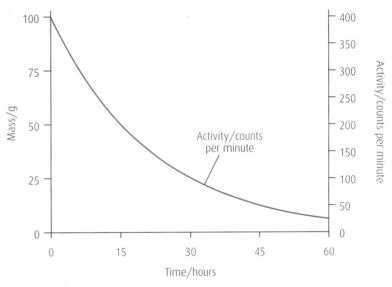

THINGS TO DO AND THINK ABOUT

Write nuclear equations for:

(a) β-decay of $^{235}_{93}U$

(b) α-decay of $^{226}_{88}Ra$

NUCLEAR CHEMISTRY: HALF-LIFE CALCULATIONS AND CARBON DATING

HALF-LIFE CALCULATIONS

EXAMPLE

Lead-210 decays to 6.25% of its original radioactivity in 84 years. What is the half-life of this isotope?

After 1 half-life, the radioactivity would have decreased from 100% to 50%.

After 2 half-lives, the radioactivity would have decreased from 50% to 25%.

After 3 half-lives, the radioactivity would have decreased from 25% to 12.5%.

After 4 half-lives, the radioactivity would have decreased from 12.5% to 6.25%.

Therefore in 84 years, 4 half-lives have passed.

$t_{1/2} = \frac{84}{4}$

$t_{1/2} = 21$ years

EXAMPLE

Astatine-210 has a $t_{1/2}$ of 8.3 hours. How long would it take for the activity of the sample to decay to $\frac{1}{32}$ of its original value?

After 1 half-life the activity would be $\frac{1}{2}$ of its original value.

After 2 half-lives the activity would be $\frac{1}{4}$ of its original value.

After 3 half-lives the activity would be $\frac{1}{8}$ of its original value.

After 4 half-lives the activity would be $\frac{1}{16}$ of its original value.

After 5 half-lives the activity would be $\frac{1}{32}$ of its original value.

5 half-lives have passed.

Therefore the time taken for the activity of the sample to decay to $\frac{1}{32}$ of its original value is $5 \times 8.3 = 41.5$ hours.

EXAMPLE

Radon-228 is a radioactive isotope found in rock, soil and water samples. It has a half-life of 5.8 years and decays by beta emission. After 17.4 years, a sample of Ra-228 had a mass of 0.05 g. What was the mass of the original sample?

First we need to work out how many half-lives have passed.

$t_{1/2} = 5.8$ years

$\frac{17.4 \text{ years}}{5.8 \text{ years}} = 3$ half-lives

After 3 half-lives (17.4 years) the mass of the Ra-228 sample = 0.05 g

After 2 half-lives (11.6 years) the mass of the Ra-228 sample = 0.10 g

After 1 half-life (5.8 years) the mass of the Ra-228 sample = 0.20 g

After 0 half-lives (0 years) the mass of the Ra-228 sample before it decayed = 0.40 g

CARBON DATING

Carbon-14 is a naturally occurring radioactive isotope of carbon. Its main use is in dating archaeological **artefacts**.

Carbon-14 is produced naturally in the upper atmosphere when nitrogen gas is bombarded with neutrons found in cosmic rays:

$$^{14}_{7}N + ^{1}_{0}n \longrightarrow ^{14}_{6}C + ^{1}_{1}p$$

The amount of carbon-14 in the atmosphere has not changed in thousands of years. Even though it decays, new carbon-14 is always being formed when cosmic rays hit atoms high in the atmosphere.

Plants absorb carbon dioxide from the atmosphere through the process of photosynthesis; animals eat plants and release carbon dioxide through the process of respiration. This means that all living things contain carbon-14.

When a living organism, like a tree, dies it stops absorbing carbon dioxide.

Carbon-14 has a half-life of 5730 years and undergoes beta decay. Scientists can use this information to determine the age of fossils or other dead organic matter by comparing how much C-14 there is in the dead organisms with that found in a living sample.

Carbon-14 is considered one of the most reliable means of determining the age of artefacts containing plant or animal matter, including some prehistoric materials up to 50 000 years old.

VIDEO LINK

For more on carbon dating, watch the video at www.brightredbooks.net/N5Chemistry.

DON'T FORGET

Carbon-14 $t_{\frac{1}{2}}$ = 5730 years.

> **EXAMPLE**
>
> **A sample of wood taken from a wooden bowl uncovered during an archaeological dig was found to contain $\frac{1}{8}$ of the carbon-14 sample taken from living wood.**
>
> Calculate the age of the wooden bowl.
> After 1 half-life the activity would be $\frac{1}{2}$ of its original value.
> After 2 half-lives the activity would be $\frac{1}{4}$ of its original value.
> After 3 half-lives the activity would be $\frac{1}{8}$ of its original value.
> Therefore 3 half-lives have passed.
> The age of the wooden bowl is 3 × 5730 years = 17 190 years old.

Geological dating

Half-life can also be used to in **geology** to date rocks. The radioisotope uranium-238 is used for dating rocks. Uranium-238 has a half-life of 4.5 billion years and decays to the stable isotope lead-206. The ratio of uranium-238 to lead-206 present in a rock can be used to determine its age.

ONLINE TEST

How well have you learned about activity of calculating half-lives and carbon dating? Test yourself online at www.brightredbooks.net/N5Chemistry.

THINGS TO DO AND THINK ABOUT

1   Which nuclei will be formed when an atom of $^{211}_{83}Bi$ emits an alpha-particle and then the product of the alpha decay loses a beta-particle?

2   A radioactive substance has a half-life of 2 hours. What fraction of the substance will remain after 4 hours have passed?

3   The half-life of cobalt-60 is 5 years. If you have 20 g of Co-60, how much will you have after 15 years?

4   Americium-242 has a half-life of 16 hours. If you started with 24 g and you now have 3 g, how much time has passed?

NUCLEAR CHEMISTRY: RADIOISOTOPES AND THEIR USES

DON'T FORGET

Radioisotopes are radioactive isotopes of elements.

USES OF RADIOISOTOPES

Medical use

Nuclear medicine is the branch of medicine that uses radiation and radioisotopes for diagnostic or treatment purposes.

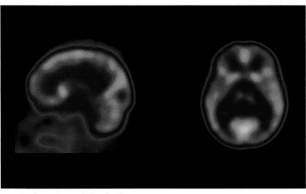

Radioisotopes are used in the production of images within the body. The radioisotope is given to patients in the form of an element or a compound. It builds up in the tissues that are being investigated. As the radioisotope decays it emits radiation, which can be traced and detected from outside the body and turned into an image that can be analysed.

Internal images of the body are taken using radioisotopes and a gamma camera

The radioisotope most widely used in diagnostic nuclear medicine as a medical tracer is technetium-99m (the 'm' indicates it is in a **metastable** form). It has ideal characteristics for medicinal use:

- Te-99m is very versatile and can be incorporated into many different compounds to form tracers that can be used in many organs and glands inside the human body.

- It has a short half-life of 6 hours, which is long enough to allow for sufficient examination and imaging but yet is short enough to minimise the dose of radiation to the patient.

- Te-99m decays by emitting gamma rays and low-energy electrons. The low-energy gamma rays it emits easily escape the human body and can be detected accurately by a gamma camera.

Iodine-123 is a radioisotope used to diagnose medical conditions affecting the **thyroid gland**. It is given to patients through a solution of the compound sodium iodide, either orally as a drink or as an injection directly into the bloodstream. It has a short half-life of 13 hours and like technetium-99m it emits gamma radiation.

Sodium chloride containing sodium-24 can be injected into the bloodstream to investigate blood flow. The beta particles emitted by the sodium-24 are tracked and monitored as the blood circulates the body. This allows the immediate detection of any circulation problems that may exist.

A thallium-201 compound can be injected into the bloodstream, where it becomes concentrated in normal heart muscle tissue but will not remain in damaged tissue. Images can be taken to locate the damaged tissue.

Radiation exposure can cause cancer but it is also used in medicine to treat cancer.

Iodine-131 is used to treat thyroid cancer and has a half-life of 8 days.

DON'T FORGET

When using radioisotopes as a diagnostic tool, a very short half-life is desirable. When they are used to treat cancers and tumours, a longer half-life is required so that treatment by the radiation produced continues over a period of time.

Cobalt-60 is used to treat tumours and is a gamma emitter. Gamma rays can be targeted at the tumour, where they destroy the cancerous tumour cells and shrink the tumour. This type of radiation is used because it can penetrate deep into the body directly to the tumour. Cobalt-60 has a half-life of 5 years, resulting in a treatment that can remain active for a long period of time.

As scientists continue to search for a cure for cancer, new radioisotopes are being developed. Two newer radioisotopes are lutetium-177 and terbium-161, which are both low-energy beta emitters with half-lives of about 6.5 days. Their short-range effect should make them useful for treating smaller tumours.

contd

INDUSTRIAL USE

Radioisotopes have a wide range of industrial applications.

Gamma irradiation

Surgical instruments are sterilised using high doses of gamma radiation from radioactive cobalt-60. Food can also be sterilised by gamma radiation in the same way. The radiation kills microorganisms, preserving the food for longer.

Monitoring the thickness of materials

Strontium-90 can be used in beta detectors that monitor and control the thickness of materials such as paper, plastic and aluminium. If the material gets thicker it absorbs more radiation and therefore less radiation reaches the detector. The detector then sends signals to the equipment, which adjusts the thickness of the material.

Identifying leaks or blockages in underground pipes

A tiny amount of radioactive material that emits gamma rays is put into the pipe. A detector is moved along the ground above the pipe. The reading on the detector increases at the leak in the pipe due to escaping radiation, which accumulates in the ground around the leak. Gamma rays can easily penetrate pipes (even if they are underground) and reach the detector. Alpha and beta particles don't have the penetrating power to pass through pipes, so cannot be used.

VIDEO LINK

Learn more by following the 'Radioactive substances' link at www.brightredbooks.net/N5Chemistry.

SCIENTIFIC USES

In addition to carbon-14 dating covered in the previous section, radioisotopes are used in many areas of scientific research:

- Tracers for monitoring physical, chemical and biological processes are spiked with radioisotopes, which can function alongside the stable elements and do not affect the overall system.

- Chlorine-36 is used to measure sources of chlorides and in dating water samples.

- Lead-210 has been used to date layers of sand and soil up to 80 years old.

- Cobalt-60, lanthanum-140, scandium-46, silver-110m and gold-198 have been used together in **blast furnaces** to monitor the **yield** and measure the performance of the furnace.

- Gold-198 and tellurium-99m can be used as tracers added to sewage so it can be monitored in the environment.

- Carbon dioxide containing carbon-14 has been used to investigate the process of photosynthesis.

ONLINE

Look inside a smoke alarm and learn more by checking out the link at www.brightredbooks.net/N5Chemistry.

DON'T FORGET

Alpha radiation is used because it has low penetrating power.

SMOKE ALARMS

Smoke alarms found commonly in the home use the radioisotope americium-241, which is a source of alpha particles with a half-life of 432 years.

A minute amount of americium-241 (approximately $\frac{1}{5000}$ th of a gram) is contained in an ionisation chamber, which is made up of two metal plates connected to a battery. The battery applies a voltage to the plates, charging one plate positive and the other negative.

As air passes through the chamber, alpha particles emitted by the americium-241 knock electrons off atoms in the air, ionising the oxygen and nitrogen. The positively charged oxygen and nitrogen ions are attracted to the negative plate and the electrons to the positive plate, generating a small electric current.

If smoke enters the chamber, particles attach to the ions, neutralising them and stopping them reaching the charged plates. A drop in current occurs between the plates, triggering the alarm.

THINGS TO DO AND THINK ABOUT

Carry out your own research to find even more examples of uses of radioisotopes.

EVERYDAY LABORATORY SKILLS

One of the important parts of this course is the development of key scientific skills. As well as important practical skills, you should be given opportunities to develop problem-solving and data-handling skills. Practical skills are an important part of any science subject. In chemistry there are some basic practical techniques that are regularly used and so it is important for you to be able to carry them out correctly. Not all of the practical skills outlined in the course are dealt with here. In this guide, you have already encountered titration, methods for following rates of reaction, electrical conductivity and cells, and determining energy from fuels.

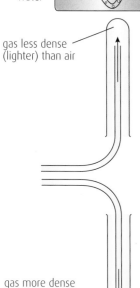

gas of low water solubility

water

gas less dense (lighter) than air

gas more dense (heavier) than air

VIDEO LINK

Find out more by watching the video 'Collecting a gas over water' at www.brightredbooks.net/N5Chemistry.

COLLECTING A GAS

Collecting a gas produced in a chemical reaction is a useful technique. It can allow the gas products to be identified and the volume of the gas to be measured. This allows information to be calculated about the reaction quantities. There are different techniques used to collect a gas depending on the properties of the gas.

Collecting a gas over water

This method is one of the most common used in school labs and works well for gases that are not very soluble in water such as carbon dioxide and hydrogen. It is also often used to collect oxygen.

The gas produced in the reaction travels through the delivery tube, where it rises to the top of the test tube. This causes the water in the test tube to be pushed down and out at the bottom.

Collecting gases that are less dense than air – downward displacement of air

This method allows gases that are less dense than air (e.g. ammonia) to be collected, as well as providing a way to collect dry hydrogen.

The gas produced in the reaction travels through the delivery tube, where it rises to the top of the test tube. This causes the air in the test tube to be pushed down and out at the bottom.

Collecting gases that are more dense than air – upward displacement of air

This method allows gases that are more dense than air (e.g. carbon dioxide and chlorine) to be collected.

The gas produced in the reaction travels through the delivery tube where it sinks to the bottom of the test tube. This causes the air in the test tube to be pushed up and out of the top.

Measuring the volume of gas produced

In some experiments it may be necessary to measure the volume of gas produced in the reaction. This can be useful when measuring rate of reaction.

This can be done with a gas syringe; the gas produced in the reaction pushes the plunger, and the volume of collected gas can be measured by reading the position of the plunger against the graduations.

100 cm³ measuring cylinder

A more common method in the school lab is to use an upturned measuring cylinder and displace the water from the measuring cylinder.

contd

TESTING GASES

Testing for gases is an important technique to learn in chemistry. A chemical test is a procedure that can be used to prove the identity of a substance. This is because a chemical test will only give a certain result for that one substance. There are a number of gas tests that you need to know about.

Testing for oxygen – Oxygen gas will relight a glowing wooden splint.

Testing for hydrogen – Hydrogen gas burns with a squeaky pop.

Testing for carbon dioxide – To test for carbon dioxide, the test gas is mixed with limewater (calcium hydroxide solution). Limewater turns cloudy (or milky) when carbon dioxide is present.

DON'T FORGET

The condenser consists of an inner tube through which the condensing gas passes, and an outer jacket. The outer jacket is connected to a tap (water is passed in at the bottom of the condenser and out at the top) and a small flow of cold water is passed through the jacket. This keeps the condenser cold and allows the gases to be condensed.

DISTILLATION

Distillation is a common technique used to separate a mixture when at least one of the substances in the mixture is a liquid. In the diagram, the mixture being separated is ethanol and water.

Ethanol and water have different boiling temperatures. The boiling point of ethanol is 78°C and that of water is 100°C. Heat is supplied to the mixture until the ethanol begins to boil and turns into a gas (**evaporate**). As the ethanol vapour passes the thermometer, the temperature measured will correspond to the temperature of ethanol vapour (around 78°C). The gas then enters the **condenser** and **condenses** back into a liquid, drips down the condenser and collects in the conical flask. This liquid is called the **distillate**. In the experiment opposite, the distillate will be ethanol.

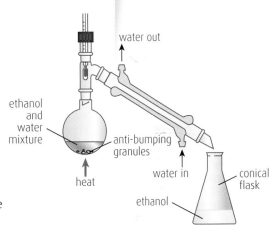

Once the water begins to evaporate the temperature on the thermometer will increase to show the temperature of water vapour (100°C). This indicates that all of the ethanol has been removed and water vapour is entering the condenser. The heat source should be removed at this point and a new collecting flask needs to be put in place in order to collect the water that is condensing.

Whisky stills are usually made from copper (laboratory stills are glass) and different distilleries use different shapes of copper still.

When distilling a mixture of ethanol and water, it is not safe to use a Bunsen burner as ethanol is flammable. A heating mantle is often used for this purpose. An electric hotplate with an oil or water bath could also be used.

ONLINE

Learn more about how whiskey is made by following the link at www.brightredbooks.net/N5Chemistry.

ONLINE TEST

Check out your chemistry skills by taking the test at www.brightredbooks.net/N5Chemistry.

METHODS OF HEATING

In the laboratory, there are a number of methods that can be used to provide heat to a process or reaction.

The most common method of heating uses a Bunsen burner; this provides a quick and easy to control source of heating and is suitable for most school experiments.

When flammable substances are being used it is not advisable to use a Bunsen burner as the flame can cause the substance to ignite. In cases like these, sometimes boiling water from a kettle will provide enough heat.

In the distillation of ethanol, a **heating mantle** may be used to provide the heat needed to boil the mixture. A heating mantle is an electric heater specially designed to hold the round-bottomed flask holding the ethanol and water mixture. In this case, it is not safe to use a Bunsen burner as ethanol is flammable.

An alternative method of heating a flammable substance is to use an electric hot-plate. This is similar to an electric hob that you may have on your cooker at home.

STANDARD SOLUTIONS

A **standard solution** is a solution with an accurately known concentration. It is useful to be able to make a standard solution as they can be used in analysis of other substances, for example in a titration.

Let us consider the titration of an unknown concentration of diluted hydrochloric acid with sodium carbonate solution. In order to determine the concentration of the diluted hydrochloric acid, a solution of sodium carbonate of accurately known concentration is needed. It will be necessary to make a **standard solution** of sodium carbonate.

The first step in this process is to work out the mass of sodium carbonate that is needed to make a certain volume of solution. If, for example, we wished to make 250 cm³ of a 0.1 mol l⁻¹ solution, we would need to weigh, accurately, approximately 2.6 g of sodium carbonate.

It will not matter if slightly less or slightly more than 2.6 g of sodium carbonate is measured, so long as the actual mass is recorded. The accurate, actual mass can then be used to calculate the exact concentration of the solution.

To measure approximately 2.6 g of sodium carbonate, a balance is used.

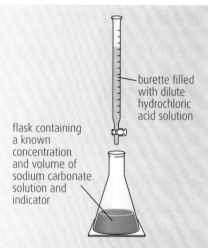

burette filled with dilute hydrochloric acid solution

flask containing a known concentration and volume of sodium carbonate solution and indicator

USING A BALANCE

A balance is an essential piece of equipment in any lab. It is used to measure mass (how much of something there is). In the lab the unit for mass is usually grams (g).

Some balances are used for measuring an approximate mass of a substance and will have only one decimal place. Others are used for more accurate measurements and may have four decimal places – these are called analytical balances.

Here are some important instructions to follow when using a balance:

* Ensure that the pan is clean. If the pan is dirty then it will not be possible to take an accurate mass measurement.

* Switch on the balance using the power button.

* Wait for the display to settle. When it is ready for use it will have a reading of zero grams.

* Gently place the object being measured onto the centre of the pan.
  Record the reading, with units, being sure to record all the decimal places.
  For example, a reading of '1.50 g' would be written down as 1.50 g and not as 1.5 g.

Weighing substances for standard solutions

To accurately weigh a substance for a standard solution the following steps are followed:

* Place a beaker onto the balance.

* Press the TARE button. This will zero the balance display and will allow the mass of sodium carbonate to be displayed without the mass of the beaker.

* Add approximately 2.6 g of sodium carbonate.

* Record the accurate mass of sodium carbonate that has been added.

MAKING THE STANDARD SOLUTION

The next step in making a standard solution of sodium carbonate is to dissolve the solid in the beaker in a small quantity of distilled water. Once the solid has dissolved, the solution is carefully poured into a standard flask. The picture below shows a standard flask. Standard flasks come in different sizes and have only one graduation mark on them.

To remove every last trace of sodium carbonate left in the beaker, the beaker needs to be rinsed with distilled water and the rinsing added to the standard flask. A small volume of water is used and the beaker will need to be rinsed two or three times, with the rinsings being added to the standard flask each time. It is important to avoid adding too much water at this stage. Ideally, the level of the solution should be below the thin neck of the flask.

Distilled water is added to the flask, carefully, until the meniscus is just below the graduation mark. A dropper is used to add water until the bottom of the meniscus is sitting on the graduation mark.

A good fitting stopper is placed in the flask and the flask is inverted (turned upside down). The bulb shape of the flask allows the solution to be mixed well.

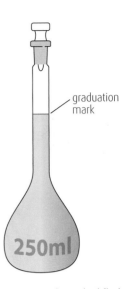

graduation mark

250ml

Filled 250 ml standard flask

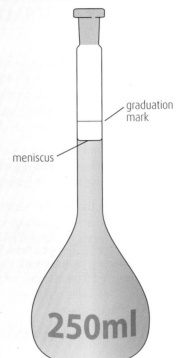

graduation mark

meniscus

250ml

THINGS TO DO AND THINK ABOUT

1 Define:

   (a) standard solution

   (b) titration

   (c) end-point

   (d) meniscus

2 Why must the rinsings from the beaker be added to the standard flask?

3 Explain the importance of having the bottom of the meniscus of the solution in line with the graduation mark on the flask.

4 In an experiment, a pupil weighs out 0·66 g of sodium carbonate to be used to make a standard solution. The instructions asked for about 0·7 g to be weighed out. Explain why it is not necessary for the pupil to try to add more to her sample of sodium carbonate.

MAKING A SALT

Salts are useful chemicals. As well as the common salt (sodium chloride) that is added to fish and chips, salts are used in fertilisers, to make plaster casts and in indigestion remedies. There are different methods that can be used to make salts. The method chosen will depend on which salt is required.

DON'T FORGET

A salt is a compound in which the hydrogen ion of an acid has been replaced by a metal ion (and sometimes ammonium). In other words, it is a product of neutralising an acid.

DON'T FORGET

The second part of a salt name comes from the acid that has been neutralised. The first part of a salt name comes from the base that was used to neutralise the acid. Check out the section 'Acids and bases: neutralisation reactions' on p48 to revise this.

MAKING A SALT BY NEUTRALISING AN ACID WITH A SOLUBLE BASE

During a titration, an acid is added from a burette to a soluble base (alkali) in a conical flask until neutralisation occurs. The end of the reaction is seen when an added indicator changes colour. The products in the conical flask will be salt and water. An example is the neutralisation of dilute hydrochloric acid using sodium hydroxide solution. The salt in this reaction is sodium chloride.

The titration is repeated until concordant volumes are obtained and an average volume is then calculated. The titration can be repeated, without indicator, using the burette to add the exact volume of base needed to neutralise the acid.

To obtain a pure sample of the salt, the water is evaporated. This can be carried out using a Bunsen burner and an evaporating basin. A Bunsen burner is a safe method to use in this experiment as none of the reactants or products is flammable.

Care is taken during heating to make sure that not all the water is removed from the evaporating basin. The small volume of water remaining will evaporate naturally from the basin.

The salt will be left in the evaporating basin.

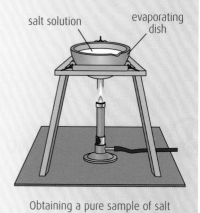

salt solution evaporating dish

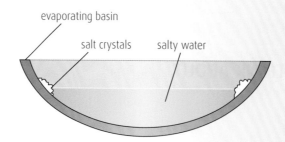

evaporating basin
salt crystals salty water

Obtaining a pure sample of salt

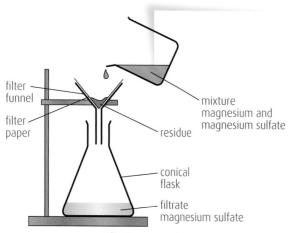

filter funnel
filter paper
mixture magnesium and magnesium sulfate
residue
conical flask
filtrate magnesium sulfate

Filtration apparatus

MAKING A SALT BY REACTING AN ACID WITH A METAL OR AN INSOLUBLE BASE

Salts can also be produced by adding a reactive metal to an acid. For example, magnesium sulfate can be made by adding magnesium to dilute sulfuric acid. This reaction also produces hydrogen gas:

$$Mg(s) + H_2SO_4(aq) \rightarrow MgSO_4(aq) + H_2(g)$$

The acid will be completely reacted when no more gas is produced and unreacted magnesium is left in the beaker. To isolate the salt, the unreacted magnesium must first be removed by **filtration**.

contd

The solid collected in the filter paper is called the **residue**. In this example the residue is unreacted magnesium. The liquid or solution collecting in the conical flask is the **filtrate**. In this example the filtrate is the magnesium sulfate solution.

The magnesium sulfate solution can then be poured into an evaporating basin and the water evaporated using a Bunsen burner.

This same process could be carried out using an insoluble salt such as magnesium carbonate or magnesium oxide.

MAKING AN INSOLUBLE SALT

The salts that were made in the previous experiments were all soluble in water and to obtain the solid salt the water was evaporated. Insoluble salts such as silver(I) chloride and lead(II) iodide can be made by reacting two soluble salts together. If we wanted to make pure silver(I) chloride we could do this by reacting a solution of silver(I) nitrate and a solution of sodium chloride.

Step 1

a solution of silver(I) nitrate is added to a solution of sodium chloride

Step 2

the precipitate is filtered off

Step 3

the filtered precipitate is washed several times with deionised (pure) water

Step 4

the silver(I) chloride is carefully scraped off the filter paper into a dish and dried in an oven

filter paper

filter funnel

precipitate

filtrate

the purified silver(I) chloride

The precipitation process

This process is known as **precipitation**. The insoluble salt formed in the reaction is removed by **filtration**.

VIDEO LINK

Watch the 'Separating mixtures - filtration and distillation' video at www. brightredbooks.net/ N5Chemistry.

ONLINE TEST

Take the test 'Salts, naming and balanced equations from metals and insoluble bases' at www.brightredbooks.net/ N5Chemistry.

THINGS TO DO AND THINK ABOUT

1. A pure sample of salt is required to be produced from the following reactions. Decide which method would be most appropriate to prepare a sample of the following salts:

   (a) copper(II) sulfate from copper(II) carbonate and sulfuric acid

   (b) potassium nitrate from potassium hydroxide and nitric acid

   (c) barium sulfate from barium nitrate and sodium sulfate.

2. Define:

   (a) filtration

   (b) filtrate

   (c) residue

3. Explain how the water can be removed from a solution of a salt made during a titration reaction.

TABLES, BAR CHARTS AND GRAPHS 1

Well-presented tables, charts and graphs provide a concise summary of experimental data. In general, tables are useful for giving actual numerical information and bar charts and graphs are useful for indicating trends, making comparisons and showing relationships.

One of the most important skills you will learn in science subjects is how to construct tables, bar charts and graphs in order to best display the results from experiments.

CONSTRUCTING A TABLE OF EXPERIMENTAL DATA

Tables are used frequently in chemistry. They may be used to display general information about elements or compounds. An example is shown below.

Alkene	Full structural formula
ethene	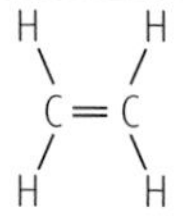
propene	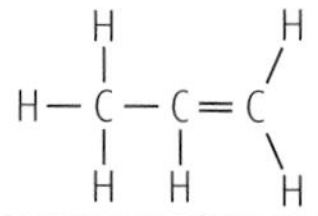

Tables are also extremely useful for summarising data from experiments. In this section we will focus mainly on tables of experimental data, although the same principles for constructing tables apply to both types.

There is rarely a 'correct' method for constructing a table of experimental data but there are some guidelines that apply to all table designs.

- Tables look best if they have a top, bottom and sides.

Table heading 1	Table heading 2

Table heading 1	Table heading 2
		✔

- Units should be included in the headings and not listed with each entry in the table.

Time	Volume
2 min	15 cm³
4 min	23 cm³
6 min	30 cm³

Time (mins)	Volume (cm³)
2	15
4	23
6	30	✔

- Select the best unit for the data.

Sometimes the data you have may contain a mixture of units. A common example is when recording time. The experiment may be collecting data every 10 seconds and if the experiment lasts for several minutes it may be that the times noted are in a mixture of seconds and minutes. It is important to select an appropriate unit – in this case either seconds or minutes. Only one should be used.

DON'T FORGET

Straight lines should always be drawn using a ruler. It helps to use a pencil when drawing a table so that mistakes can be easily corrected.

DON'T FORGET

Time is not a decimal scale and so 1 minute and 30 seconds would be 1·5 minutes and not 1·30 minutes.

CONSTRUCTING A BAR CHART

Bar charts and graphs are an incredibly useful tool to a scientist as they allow data to be displayed and understood in a quick and easy manner. They allow generalisations to be made and conclusions to be drawn for experiments.

A bar graph is commonly used when only one set of experimental data involves numbers. For example, if an experiment was carried out to measure the different energy values of various fuels, this information would be represented by a bar chart. It is common to have the numerical values plotted on the *y*-axis.

Look at the bar chart shown.

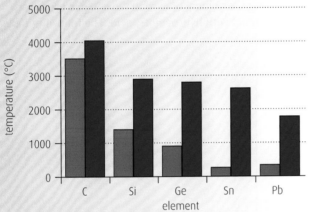

Melting and boiling points of group 4 elements

This bar chart shows the different group 4 elements on the *x*-axis with the numerical values for their melting and boiling points plotted on the *y*-axis.

Here are some general guidelines:

- The scale for the *y*-axis needs to be chosen carefully. It needs to be linear, for example 0, 1000, 2000, 3000, 4000 and not 0, 500, 1000, 2000, 4000.

- The graph should be as big as possible and should certainly use at least half of the graph paper provided.

- Bars should be drawn using a pencil and ruler and should be of equal width.

- Both axes need to be carefully labelled, using units as appropriate.

ONLINE

Make your own bar graph at www.brightredbooks.net/N5Chemistry.

ONLINE TEST

Take the 'Tables and Graphs' test online at www.brightredbooks.net/N5Chemistry.

DON'T FORGET

A linear scale is one in which each division on a graph represents the same numerical value. In this example, the first division on the *y*-axis represents 1000°C. This means that all divisions must also be 1000°C.

THINGS TO DO AND THINK ABOUT

1 Suggest three improvements that could be made to the table below.

Acid used	Trial 1 Volume collected	Trial 2 Volume collected
hydrochloric acid	500 cm³	520 cm³
sulfuric acid	990 cm³	1 litre

2 The table shows the results for an experiment to find out which metal, when paired with zinc, gave the highest voltage in an electrochemical cell. Construct a bar chart containing these data.

Electrode	Voltage (V)
iron	0·5
lead	1·2
copper	1·8
magnesium	0·8

TABLES, BAR CHARTS AND GRAPHS 2

CONSTRUCTING A GRAPH

Graphs are used to present data when both sets of experimental data involve numbers and these numbers are continuous.

Generalisations and conclusions can only be made from graphs if they have been correctly drawn.

We shall use the example of measuring the volume of carbon dioxide gas produced every 30 seconds when marble chips react with hydrochloric acid. The following table of data was produced:

Time (s)	Volume (cm³)
0	0
30	12
60	32
90	51
120	64
150	73
180	80
210	81
240	81

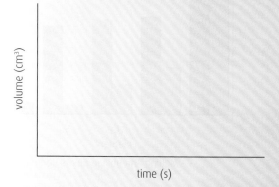

In most cases, the variable you are controlling is plotted along the x-axis (bottom). In this case the variable being controlled is time – we are choosing to measure the volume every 30 seconds. Volume would be plotted on the y-axis.

Scales for both the x- and y-axes need to be chosen carefully.

 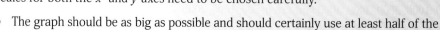
- The graph should be as big as possible and should certainly use at least half of the graph paper provided.

- The scales chosen need to be linear. For example, if one box on a graph is worth 10 cm³, then each box on that graph must equal 10 cm³. This may mean that sometimes a data point will be plotted between two divisions (if, for example, 15 cm³ was being plotted, this would come between the 10 cm³ and 20 cm³ divisions).

Using the graph paper below, scales can be drawn as shown.

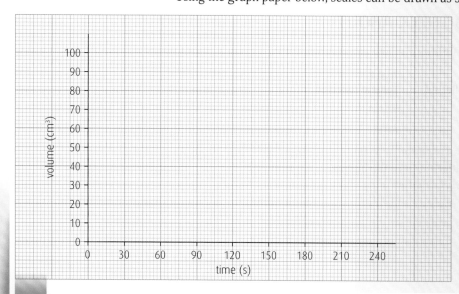

- The data points are plotted carefully using small marks – crosses are commonly used but small dots are also acceptable.

- You should also consider if your graph should be going through the origin. This will depend on the experiment. In the experiment above, at the start of the reaction the volume of gas would be zero and so when drawing the line it should go through the origin.

Your graph should now be ready for a line of best fit to be drawn.

DRAWING A LINE OF BEST FIT

When drawing a line on your graph it is necessary to first of all consider what your experimental data are. Mostly, the data you have plotted on your graph will have been from measuring a quantity in a reaction. There will be errors associated with this value (is it easy to read an exact volume measurement of a gas at exactly 30 seconds?). As such the value you recorded ($12\,cm^3$) will not be exactly this value and could be a little less or a little more than this value. For this reason, lines of best fit are drawn. This shows a general trend line that goes through, or as close to, as many data points as possible.

There are two main types of line of best fit.

Best-fit straight line

This is probably the easiest to draw. If your experimental data look to be in a straight line, then, using a ruler, a straight line can be drawn that takes in as many of the data points as possible. This is done 'by eye', and an easy way to do it is to try to have as many points below the line as above it.

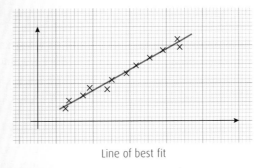

Line of best fit

Best-fit curve

This is slightly trickier as the line will need to be drawn free-hand. Again, the line should go through, or as close to, as many of the data points as possible and you should aim to have as many points below the line as above. The line needs to be smooth and sharp (do not try to sketch the line but draw it using one continuous hand movement).

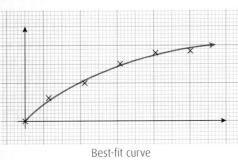

Best-fit curve

The experiment outlined on page 110 will produce a similar-shaped graph to the best-fit curve shown here.

ONLINE

Check out the 'Scatter diagrams' link at www.brightredbooks.net/N5Chemistry.

ONLINE TEST

Take the 'Tables and graphs' test at www.brightredbooks.net/N5Chemistry.

DON'T FORGET

When adding a trend line to your graph you will need to decide which type of line best suits the data you have plotted.

THINGS TO DO AND THINK ABOUT

1   For the following examples, decide which variable should be plotted along the *x*-axis.

   (a) In an experiment to determine the rate of a reaction, the mass of the flask the reaction took place in was recorded every 2 minutes.

   (b) A reaction was carried out at 10, 20, 30 and 40°C and the time taken for the reaction was recorded.

2   Look at this graph and explain what is wrong.

3   For the following examples, decide if the line of best fit should go through the origin.

   (a) A reaction was carried out at 10, 20, 30 and 40°C and the time taken for the reaction was recorded.

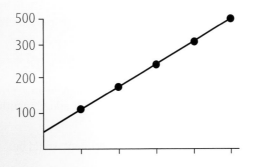

   (b) A reaction was carried out at 0·1, 0·2, 0·3 and 0·4 mol l⁻¹ and the time taken for each reaction was recorded.

ANALYSING DATA FROM TABLES, BAR CHARTS AND GRAPHS 1

As well as summarising experimental data, tables, bar charts and graphs can be used to predict values for experiments and can also provide general trends, allow comparisons to be made and show the relationship between two variables. Being able to analyse data in this way is an extremely useful skill.

ANALYSING DATA FROM A TABLE

Tables of experimental data provide us with fixed numerical values for an experimental variable. Take a look at the table of energy values for different fuels shown below.

Name of fuel	Number of carbons	Energy released when one mole of fuel burns (kJ)
methanol	1	726
ethanol	2	1367
propanol	3	2017
butanol	4	2665

The information in this table can be used in two ways.

1. To make a statement about the general trend.

In this example we can see that there is an increase in the energy released as the number of carbon atoms in the fuel increases.

2. To predict a value for the energy released for a fuel not given in the table.

You may be required to predict a value for another fuel, for example pentanol, which contains five carbon atoms per molecule. To work this out it is necessary to see if there is a pattern in the energy values as the number of carbon atoms increases.

Number of carbons	Energy released when one mole of fuel burns (kJ)
1	726
2	1367
3	2017
4	2665

}641
}648
}650

These values are the differences in energy between the fuels. For example, 641 is the difference between the fuel with two carbons and the fuel with one carbon (1367 − 726)

The average increase in energy per carbon atom added is around 646 kJ $\left(\frac{641 + 648 + 650}{3}\right)$.

2665 + 646 = 3311 kJ

The predicted energy released when one mole of pentanol (five carbons) burns is therefore 3311 kJ.

CALCULATING AVERAGES (MEANS)

Often a table is used to show data from repeat experiments.

For example, a student carried out an experiment to find the time taken for a reaction to finish. The experiment was repeated another two times. The following results were obtained.

	Experiment 1	Experiment 2	Experiment 3
Time taken (seconds)	320	315	322

To use this data, it is necessary to calculate an average (mean) of all three results. To calculate an average, all the values are added together and the total is divided by the number of sets of values.

So, the average time for this experiment would be:

$\frac{320 + 315 + 322}{3}$ =319 seconds

Any repeat experiments which give results that are very different from the others can be disregarded as being rogue data because something went wrong during the experiment.

ANALYSING DATA FROM A BAR CHART

Bar charts can be analysed in a similar way to tables.

Consider this bar chart. It shows the energy needed to remove an electron from a mole of atoms for the first 19 elements of the Periodic Table.

At first glance, this bar chart is confusing as it contains a lot of data. The information can be broken down into smaller chunks. The bars are plotted for atomic number and three elements are highlighted: helium, neon and argon. All of these elements are in the same group in the Periodic Table (group 8). Closer inspection shows that elements 3–10 are in one row (period) of the Periodic Table and elements 11–18 are in another.

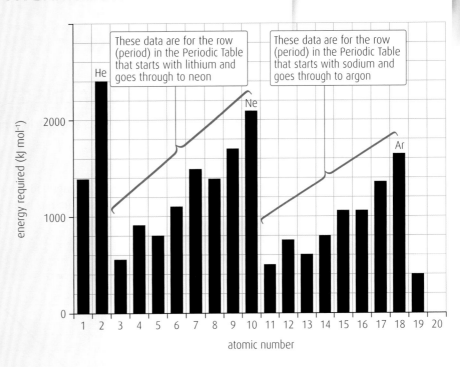

These data are for the row (period) in the Periodic Table that starts with lithium and goes through to neon

These data are for the row (period) in the Periodic Table that starts with sodium and goes through to argon

Now that this information has been worked out, it is possible to pick out general trends. The first trend concerns a row in the Periodic Table. The general trend for the elements in the row beginning with atomic number 3 (lithium) is that as the atomic number increases so too does the energy required to remove an electron. This trend is also true for the next period (staring with atomic number 11 (sodium)). You will see in both of these trends that there are exceptions. Elements with atomic numbers 4 and 12 have higher energies than the element next to them. These exceptions can be ignored as it is only the general trend that is being looked at.

Further examination allows us to gain information about elements within groups of the Periodic Table. In moving down the elements in group 8 (He, Ne and Ar) the energy values decrease. This is also true for other groups – elements 3, 11 and 19 all show a similar trend and are all elements in group 1 of the Periodic Table.

Putting all of this information together allows a prediction to be made as to the energy value of the element with atomic number 20. It must be a higher energy value than that for element 19 (increasing energy as you move along a period) and must be lower energy value than that for element 12 (decrease in energy as you move down a group). A possible value of 600 kJ would fit with these trends.

 ONLINE TEST

Take the 'Tables and graphs' test online at www.brightredbooks.net/N5Chemistry.

THINGS TO DO AND THINK ABOUT

The table below shows the relationship between the solubility of sulfur dioxide in water and the temperature of the water.

Temperature (°C)	0	10	20	30	40	60	80
Solubility (grams per litre)	225	145	95	60	35	15	5

(a) Draw a line graph of solubility against temperature.

(b) State the relationship between the solubility of sulfur dioxide in water and the temperature of the water.

ANALYSING DATA FROM TABLES, BAR CHARTS AND GRAPHS 2

ANALYSING DATA FROM A LINE GRAPH

Graphs provide us with general trends and conclusions from experiments as well as allowing values to be predicted.

General trends and conclusions

The table below is from the example on page 110 using marble chips and dilute hydrochloric acid. The graph for these data has been plotted.

Time (s)	Volume (cm³)
0	0
30	12
60	32
90	51
120	64
150	73
180	80
210	81
240	81

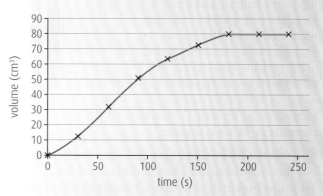

From the graph it is clear to see that the general trend is that as the time increases, so too does the volume of gas produced.

We can also see that a larger volume of gas is produced in a given time period in the initial stages of the reaction than is produced in the same time interval towards the end of the experiment, for example 20 cm³ of gas is produced in the 30 seconds between 30 and 60 seconds and only 9 cm³ of gas is produced in the 30 seconds between 120 and 150 seconds.

Data from a graph

As well as allowing us to make conclusions and generalisations, graphs are useful for getting data.

Consider the graph shown below.

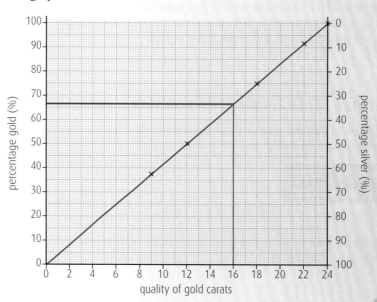

contd

The percentage gold has been determined, by experiment, for 9-, 12-, 18-, 22- and 24-carat gold. We can see that there is a line of best fit and that the data points fit well on this line. This allows information to be obtained for other carats of gold that were not tested in the original experiment. For example, a value for the percentage gold in 16-carat gold can be read from the graph.

- From the x-axis, trace a finger along the line representing 16 carats (red line).
- Use a ruler to extend a line from the line of best fit to the y-axis (blue line).
- Carefully read the scale on the y-axis.

From this graph, the percentage gold in 16-carat gold is 67%.

ONLINE TEST

Take the 'Tables and graphs' test online at www.brightredbooks.net/N5Chemistry.

What else does a graph tell us?

Another use for a graph is to assess how good the experiment was. If, in drawing a best-fit trend line, the line goes through or near most of the data points then we can assume that the experiment was reasonably reliable. Occasionally a graph will allow a 'rogue' result to be identified.

This graph has a best-fit straight line drawn in.

Looking at this graph it is clear that the data point at concentration 0·6 mol l⁻¹ is away from the line of best fit. This data point is a rogue result. As there are many data points in this experiment, the rogue result does not affect the line of best fit but in some experiments this value would need to be repeated to enable a more accurate best-fit line to be drawn.

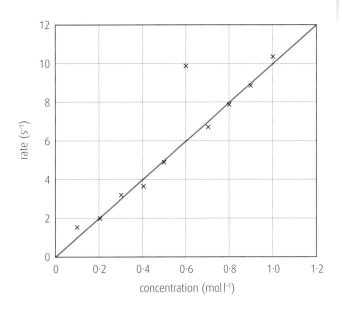

THINGS TO DO AND THINK ABOUT

Rapid inflation of airbags in cars is caused by the production of nitrogen gas. The graph gives information on the volume of gas produced over 30 microseconds.

(a) State the volume, in litres, of nitrogen gas produced at 2 microseconds.

(b) State the volume, in litres, of nitrogen gas produced at 10 microseconds.

(c) At what time was half of the final volume of nitrogen gas produced?

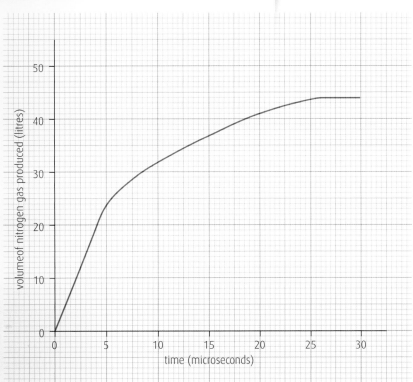

EXPERIMENTAL DESIGN

The ability to be able to design an experiment is a useful skill for a chemist to have. Without a good experiment design, the results cannot be relied upon.

FAIR TESTING

The first thing to consider when designing an investigation is: 'Is the investigation fair?'. A fair test is when only one of the factors (variables) of the experiment is altered at a time, whilst all the other variables are kept the same.

Let us consider an investigation to see how altering the electrolyte affects the voltage produced in an electrochemical cell. The diagram below shows the first experiment carried out during this investigation.

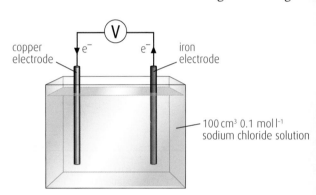

copper electrode e^- V e^- iron electrode

100 cm³ 0.1 mol l⁻¹ sodium chloride solution

To investigate what effect the electrolyte will have on the voltage produced by this cell the electrolyte will need to be changed in further experiments. This is the variable that is to be investigated. To ensure that changes in voltage are only due to the changes in the electrolyte, all other variables will need to be kept the same. This ensures that the investigation is fair. Copper and iron electrodes will need to be used, the volume and concentration of the electrolyte will need to stay the same and the temperature the experiment is carried out at will need to be constant.

DON'T FORGET

If repeat experiments do not give the same (or nearly the same) results, then it is likely that there is some error in the way the experiment is being carried out and so the experiment design would need to be altered.

REDESIGNING AN EXPERIMENT

Sometimes when we look at our results we can see that the experiment needs to be redesigned in some way.

Sometimes results from experiments can be completely unexpected and this can mean that the experiment design was wrong. This was seen in the example given on the page 115 where one point on a line graph was clearly unexpected compared with the other results.

ONLINE TEST

Take the 'Examples of experimental results and suggestions of improvements' test online at www.brightredbooks.net/N5Chemistry.

EQUIPMENT DIAGRAMS

In designing an experiment, often a diagram of the equipment is required. There are some key points to consider when drawing such diagrams. Firstly, break down the experiment into two key parts:

- reactants
- products

Reactants

Begin by considering how the reaction is going to take place:

- Does the experiment need to be heated or cooled in ice?
- Do any of the reactants need adding slowly over a period of time?

Products

Next, consider the products – how are they to be collected?

contd

Experiments involving collecting gases need to be sealed systems and the diagram must show this. Stoppers or bungs and delivery tubes are common in experiments involving gases.

When gases are being passed through test tubes, care needs to be taken to show correctly how the gas will be drawn through the equipment. This is shown in this diagram.

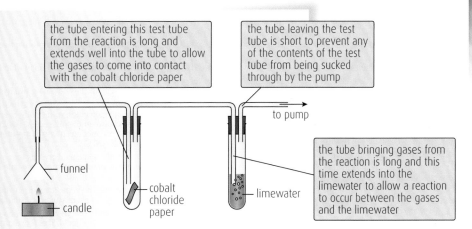

the tube entering this test tube from the reaction is long and extends well into the tube to allow the gases to come into contact with the cobalt chloride paper

the tube leaving the test tube is short to prevent any of the contents of the test tube from being sucked through by the pump

to pump

the tube bringing gases from the reaction is long and this time extends into the limewater to allow a reaction to occur between the gases and the limewater

funnel

candle

cobalt chloride paper

limewater

DRAWING EXPERIMENTAL DIAGRAMS

Chemists are often required to draw diagrams of the equipment or apparatus used in an experiment. The following guidelines should be followed:

- A ruler and pencil should be used.
- Labels are essential for equipment and chemicals.
- Mostly, line or cross-sectional diagrams are encouraged.

Cross-sectional diagram **A** is what you would see if you sliced through the apparatus with a knife. Imagine cutting through an orange with a knife. Inside are the seeds and segments, which you cannot see from the outside.

Diagram **A** is cross-sectional – we can see the delivery tube inside the stopper as if it had been sliced through the middle.

Diagram **B** is not cross-sectional.

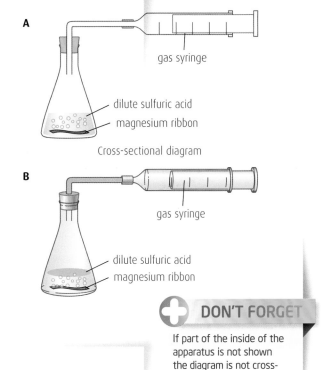

A

gas syringe

dilute sulfuric acid

magnesium ribbon

Cross-sectional diagram

B

gas syringe

dilute sulfuric acid

magnesium ribbon

> **DON'T FORGET**
>
> If part of the inside of the apparatus is not shown the diagram is not cross-sectional.

THINGS TO DO AND THINK ABOUT

1 An investigation is being designed to see how the electrolyte will affect the voltage produced by an electrochemical cell. The first experiment for this is shown in Diagram A. Copy and complete diagram B to show how the experiment would be altered to show how the voltage changes when the electrolyte is changed from sodium chloride to hydrochloric acid.

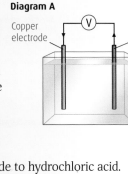

Diagram A

Copper electrode

Zinc electrode

V

150 cm³
_____ mol l⁻¹
sodium chloride

Diagram B

V

_____ cm³
_____ mol l⁻¹
hydrochloric acid

2 Eggshells are made up mainly of calcium carbonate. A pupil carried out an experiment to react eggshells with dilute hydrochloric acid. A gas was produced. Complete the diagram to show the apparatus that could have been used to measure the volume of gas produced.

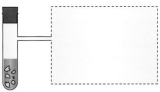

GLOSSARY

Acid – A substance that forms a solution with a pH less than 7.

Acid rain – Rain containing dissolved sulfur dioxide and/or nitrogen dioxide. It has damaging effects on buildings, structures made of iron or steel, soils, and plant and animal life.

Acidic solution – A solution that contains an excess of hydrogen ions ($H^+(aq)$).

Addition – The reaction that takes place when a molecule such as hydrogen or bromine 'adds on' across the carbon-to-carbon double bond in an alkene.

Alcohols – A family of compounds derived from hydrocarbons in which one of the hydrogen atoms is replaced with a hydroxyl (OH) group.

Alkali – A soluble base that forms a solution with a pH greater than 7.

Alkali metals – The family of reactive metals in group 1 of the Periodic Table.

Alkaline solution – A solution that contains an excess of hydroxide ions ($OH^-(aq)$).

Alkanes – A family of saturated hydrocarbons that have the general formula C_nH_{2n+2}.

Alkenes – A family of unsaturated hydrocarbons that have the general formula C_nH_{2n}. They all contain a carbon-to-carbon double bond.

Alloy – Metallic material made from a mixture of metals with other metals and/or non-metals.

Alpha particles – Helium nuclei – each containing two protons and two neutrons.

Aqueous – This refers to a solution in which water is the solvent.

Artefact – A man-made object that is of cultural or historical interest.

Atomic number – The number of protons in an atom of an element.

Atoms – The tiny particles that make up elements. An atom consists of a positively charged nucleus with negatively charged electrons moving around outside the nucleus. It has no overall charge.

Balanced formula equation – A formula equation that includes the numerical coefficients required to ensure that the number and type of atom on each side of the equation is the same for example $2H_2(g) + O_2(g) \rightarrow 2H_2O(l)$.

Base – A substance that neutralises an acid. Bases are metal oxides, metal hydroxides, metal carbonates and ammonia. Those bases that are soluble in water form alkalis.

Beta particles – High-energy electrons emitted from the nucleus.

Biodiesel – Fuel made from vegetable oils or animal fat rather than from crude oil.

Blast furnace – Industrial process used to extract iron metal from its ore.

Burning – A chemical reaction in which a substance reacts with oxygen in the air and produces heat energy.

Carbohydrate – Organic compound containing carbon, hydrogen and oxygen.

Carboxyl – The COOH functional group found in carboxylic acids.

Carboxylic acids – A family of compounds derived from hydrocarbons in which two of the hydrogens are replaced with a carboxyl (COOH) group.

Catalyst – A substance that speeds up a chemical reaction. By the end of the reaction a catalyst is unchanged chemically and the same mass is present as there was at the start.

Ceramic – A material that is permanently changed by heating, for example clay.

Chemical formula – A shorthand way of representing a substance. It shows what elements are present in the substance and the number of atoms of each. For example, CO_2 is the formula for carbon dioxide and shows that one molecule of carbon dioxide contains one carbon atom and two oxygen atoms.

Chemical reaction – A process in which substances change to form one or more new substances.

Chemical symbol – A shorthand way of representing an element. It consists of one or two letters, for example C for carbon and Co for cobalt.

Combustion – Another word for burning.

Compost – Natural fertiliser made from decayed plant matter.

Compound – A substance in which two or more elements are chemically joined.

Concentration – The number of moles of solute per litre of solution.

Concordant – In agreement. In chemistry, titration results that are within $0 \cdot 2\,cm^3$ of each other are said to be concordant.

Condensation reaction – A chemical reaction in which a small, simple molecule is produced when larger molecules are joined together.

Condense – The change of a gas into a liquid.

Condenser – The name given to a piece of equipment that cools a gas and so causes it to change into a liquid.

Covalent bonding – A covalent bond is formed when two atoms of (usually) non-metal elements share a pair of electrons. It is the force of attraction between the negatively charged shared pair of electrons and the positively charged nuclei on either side that holds the atoms together.

Covalent network – Substances made up of giant molecules in which the atoms are joined by strong covalent bonds. They have high melting and boiling points, and do not conduct electricity in any state. Examples include diamond (C) and silicon dioxide (SiO_2).

Crude oil – A liquid fossil fuel that contains a mixture of hydrocarbons.

Cycloalkanes – A family of saturated hydrocarbons that have the general formula C_nH_{2n}. They all contain a ring of carbon atoms.

Deionise – To remove ions from.

Delocalised – Used to refer to an electron that is not 'attached' to a particular atom.

Diacid – A molecule containing two carboxyl (COOH/carboxylic acid) functional groups.

Diatomic molecule – A molecule that contains only two atoms, for example H_2 and CO.

Dilute solution – A solution in which a small amount of substance (solute) has been dissolved.

Diol – A molecule containing two hydroxyl (OH/alcohol) functional groups.

Discrete covalent molecular structure – Substances made up of small, separate molecules, with strong covalent bonds inside the molecules and weak bonds between the molecules. They have low melting and boiling points, and do not conduct electricity in any state. Examples include bromine (Br_2) and ammonia (NH_3).

Dissociate – In chemistry this term refers to compounds splitting up. For example, water molecules dissociate into hydrogen and hydroxide ions: $H_2O(l) \rightarrow H^+(aq) + OH^-(aq)$.

Distillate – The name given to the liquid that is collected after it has been separated from a mixture by distillation.

Distillation – A method of separating mixtures where at least one substance in the mixture is a liquid. This technique makes use of the different boiling points of the substances in the mixture. It is used to increase the alcohol concentration of fermented mixtures.

Ductile – Easily drawn into wires.

Electric current – A flow of charged particles. Electrons flow through metals. Ions flow through solutions and melts of ionic compounds.

Electrical conductor – A substance that allows a current of electricity to pass through it.

Electrical insulator – A substance that does not allow a current of electricity to pass through it.

Electrochemical series – A list of metals and non-metals arranged in order of how easily they will lose electrons.

Electrolysis – A process in which an ionic compound is broken up into its elements using electrical energy (electricity).

Electrolyte – An ionic compound in solution that conducts electricity.

Electron – A tiny particle in an atom that is located outside the nucleus. It has a very small mass and a charge of –1.

Electron arrangement – The arrangement of electrons in energy levels around the nucleus of an atom. The electron arrangements of some elements can be found in your data booklet.

Element – A substance that cannot be broken down into a simpler substance. All the elements are listed in the Periodic Table in your data booklet.

Energy levels (shells) – The regions in an atom where electrons are found. These levels increase in size and energy as they get further from the nucleus.

Enzyme – Biological catalyst.

Esters – A family of compounds made from alcohols and carboxylic acids that are often used as flavourings and perfumes.

Evaporate – The change of a liquid into a gas. This is sometimes called boiling.

Exothermic reaction – A reaction in which energy is released.

Fermentation – A chemical reaction in which glucose is broken down into alcohol (ethanol) and carbon dioxide. It is catalysed by enzymes present in yeast.

Fertiliser – A substance that contains the essential elements needed by plants for healthy growth.

Filtrate – The solution or liquid that collects in the conical flask after a mixture has been separated by filtration.

Formula equation – A shorthand way of describing a chemical reaction, showing the formulae of the reactants and products, for example $Mg(s) + 2HCl(aq) \rightarrow H_2(g) + MgCl_2(aq)$.

Formula mass – A mass found by adding the atomic masses of all the atoms in the formula of a substance. The atomic mass of selected elements can be found in the data booklet.

Fossil fuels – Coal, oil and natural gas, formed over millions of years from the remains of dead animals and plants.

Fraction – A mixture of hydrocarbons with boiling points within a certain range of temperature.

Fractional distillation – The process used to separate crude oil into fractions according to the different boiling points of the hydrocarbons present.

Fuel – A substance that reacts with oxygen (burns) to produce heat energy.

Full structural formula – Shows the arrangement of atoms and bonds in the molecule.

Functional group – The group of atoms in a molecule that are responsible for the chemical reactions the molecule will undergo.

Functional molecular formula – A molecular formula that shows the functional group of the molecule.

Gamma radiation – A type of electromagnetic wave radiation, similar to X-rays but of higher energy.

General formula – A formula written in such a way as to be applied to an entire homologous series. For example, the general formula for the alkanes is C_nH_{2n+2}.

Geology – The scientific study of the origin, history and structure of the Earth.

Global warming – A phenomenon caused by increasing amounts of carbon dioxide in the atmosphere, leading to more of the Sun's energy being absorbed and therefore increasing temperatures in the atmosphere.